Quick Algebra Review

MORE THAN 80 SELF-TEACHING GUIDES TEACH PRACTICAL SKILLS FROM ACCOUNTING TO ASTRONOMY, MANAGEMENT TO MICROCOMPUTERS. LOOK FOR THEM ALL AT YOUR FAVORITE BOOKSTORE.

STGs on mathematics:

Background Math for a Computer World, 2nd ed., Ashley
Business Mathematics, Locke
Business Statistics, 2nd ed., Koosis
Finite Mathematics, Rothenberg
Geometry and Trigonometry for Calculus, Selby
Linear Algebra with Computer Applications, Rothenberg
Math Shortcuts, Locke
Math Skills for the Sciences, Pearson
Practical Algebra, Selby
Quick Algebra Review, Selby
Quick Arithmetic, Carman & Carman
Quick Calculus, Kleppner & Ramsey
Statistics, 2nd ed., Koosis
Thinking Metric, 2nd ed., Gilbert & Gilbert
Using Graphs and Tables, Selby

Quick Algebra Review

PETER H. SELBY
Director, Educational Technology
MAN FACTORS, INC.
San Diego, California

A Wiley Press Book

John Wiley & Sons, Inc.
New York · Chichester · Brisbane · Toronto · Singapore

Publisher: Judy V. Wilson
Editor: Alicia Conklin
Managing Editor: Maria Colligan
Composition and Make-up: Cobb/Dunlop, Inc.

Library of Congress Cataloging in Publication Data

Selby, Peter H.
  Quick algebra review.

  (Wiley self-teaching guides)
  Includes index.
  1. Algebra. I. Title.
QA154.2.S443 1982   512.9   82-21966
ISBN 0-471-86471-4

Printed in the United States of America

20 19 18 17 16 15 14 13 12 11

To the Reader

Quick Algebra Review is intended primarily as a refresher for those who have completed the Wiley Self-Teaching Guide *Practical Algebra.* However, since it covers the topics usually found in any intermediate algebra course, it should serve equally well as a review for the reader who at some time has had either a second course in high school algebra or a first course in college algebra.

Adult learners should find this book especially helpful since the review format used will enable him/her to identify, quickly and easily, the specific algebraic concepts and methods still familiar, as well as those that are hazy and therefore need special attention. (If, of course, you find you have forgotten more than you thought and perhaps need some relearning, you probably should procure a copy of *Practical Algebra* and study there the topics with which you are having difficulty.)

Unit 1 reviews some of the similarities and differences between arithmetic and algebra. This will help you get started. Subsequent units deal with these and other topics in more detail.

To help you decide if you need to read Unit 1, turn to page 1 and you will find there a short pretest. Take this test and see how you get along. If 90 percent or more of your answers are correct, you may wish to go directly to Unit 2. Otherwise it probably would be best to start with Unit 1.

INSTRUCTIONS

Each unit begins with several pages of review items presented in tabular form. In each case, an example is given and a page reference where a fuller explanation may be found. Reference numbers correspond to review item numbers. In many cases—depending, of course, on how recently you have studied algebra and how much you recall—the review item and example will refresh your memory sufficiently. However, when you find you need further help, turn to the page indicated, where you also will find additional examples and practice problems.

Contents

Pretest

1. Express the product of the following without using multiplication signs.

   (a) $a \times b \times c$ ___*abc*___

   (b) $3 \times k \times m$ ___3km___

   (c) $\dfrac{3}{4} \times 8 \times y$ ___6y___

   (d) $0.5 \times 30 \times q \times t$ ___15qt___

   (e) $2 \times 3 \times st$ ___6st___

2. Identify the literal factors in the following expressions.

   (a) $5atk$ _____

   (b) $7k\dfrac{y}{t}$ _____

   (c) $3ab \cdot 2y$ _____

   (d) $(k)(m)(t)$ _____

   (e) $0 \cdot 3by$ _____

3. Use letters and symbols to change these word statements into algebraic equations.

   (a) The sum of one-half t and twice t equals twenty.

   $\frac{1}{2}t + 2t = 20$

   (b) Eight times a number (n) minus three times the number equals five more than four times the number. $8N - 3N = 5 + 4N$

   (c) The area (A) of a triangle is equal to one-half the base (b) times the height (h). $A = \frac{1}{2}bh$

   (d) Half of c plus twice d added to five equals eight. _____

   $\frac{1}{2}c + 2d + 5 = 8$

4. What values of the indicated letter in the denominator of each of the following expressions would result in an undefined division?

(a) $\dfrac{2}{y-4}$, $y =$ _____4_____

(b) $\dfrac{3b}{4a}$, $a =$ _____0_____

(c) $\dfrac{k}{y-x}$, $x =$ _____y_____

(d) $\dfrac{0.9ky}{7.6cd}$, c or $d =$ _____0_____

5. Use parentheses correctly (where applicable) while turning these word statements into algebraic equations.

(a) Twice the sum of c and d equals eleven. $2(c+d)=11$

(b) k times the sum of x and y equals p times the quantity z minus t. $k(x+y)=p(z-t)$

✗ (c) Three divided by one-half the quantity a plus b equals fourteen. $3 \div \frac{a}{2} + b = 14$

✗ (d) y plus the quantity b minus three equals seven times the quantity four plus c. $y+b-3=7(4)+c$

(e) Three times a number (n), divided by y times the sum of five and the number, is equal to seven. $3n / y(5+n) = 7$

6. Complete the following.

(a) $3(2+3) - 7 + \dfrac{8}{2} =$ _____

(b) $\dfrac{4+2}{3} - \dfrac{8}{4} + 3 \cdot 2 =$ _____

(c) $5 - \dfrac{9}{3} + 3(2+1) =$ _____

(d) $\dfrac{3+9}{4-1} - 4 + (8 \div 2) =$ _____

7. Complete the following, letting $x = 2$ and $y = 3$.

(a) $2(x+y) - \dfrac{3}{y-x} + 7 =$ _____14_____

(b) $\dfrac{9}{y} - \dfrac{x}{2} + \dfrac{5xy}{x+y} =$ _____

$2 + \dfrac{30}{5} = 8$

(c) $\dfrac{2xy}{4} + \dfrac{1}{2}(4y - x) =$ _____ 8

(d) $\dfrac{x^2y^3}{4} + \dfrac{xy^2}{6} + y =$ _____ 9

8. How many *terms* are in each of the following expressions?

(a) $4b + \dfrac{2}{3}cx - 3(a - b)$ _____ 3

(b) $2(c + d) - \dfrac{k}{m} + 3y$ _____ 3

(c) $c(x) + b^2c$ _____ 2

(d) $ac(y + x)$ _____ 1

9. Write the following expressions using exponents.

(a) $dd + cdd + ccca$ _____ $d^2 + cd^2 + c^3a$

(b) $mmmy - xx + px$ _____ $m^3y - x^2 + px$

(c) $\dfrac{y}{xx} + yyx - y(xy)$ _____ $\frac{y}{x^2} + y^2x - xy^2$

(d) $(c + d)(c + d)$ _____ $c^2 + cd + dc + d^2$

(e) $cd + de + ef$ _____ same

10. Put into words the meaning of these expressions.

(a) b^2c^3 _____ $b \cdot b \cdot c \cdot c \cdot c$

(b) $3a^2f$ _____ $3 \cdot a \cdot a \cdot f$

(c) 5^2x^3 _____ $5 \cdot 5 \cdot x \cdot x \cdot x$

(d) $(2y)^2$ _____ $2y \cdot 2y$

11. Simplify the following.

(a) $2a + 3b + 3a - b$ _____ $5a + 2b$

(b) $3ab + 2k - ab + 3$ _____ $2ab + 2k + 3$

(c) $2(a + b) - a + 3b$ _____ $a + 3b$

(d) $ax^2 + by + b^2 + 3ax^2$ _____ $4a^2x^4 + b^3y$

(e) $3xy + 3y^2 - 2xy + y^2$ _____ $xy + 4y^2$

ANSWERS TO PRETEST

1. (a) abc;   (b) $3km$;   (c) $6y$; (d) $15qt$;   (e) $6st$

2. (a) a, t, k;   (b) $k, y, \dfrac{1}{t}$;   (c) a, b, y;   (d) k, m, t;   (e) b, y

3. (a) $\dfrac{t}{2} + 2t = 20$;   (b) $8n - 3n = 4n + 5$;   (c) $A = \dfrac{1}{2}bh$;
   (d) $\dfrac{c}{2} + 2d + 5 = 8$

4. (a) $y = 4$;   (b) $a = 0$;   (c) $x = y$;   (d) c or $d = 0$

5. (a) $2(c + d) = 11$;   (b) $k(x + y) = p(z - t)$;   (c) $\dfrac{3}{\frac{1}{2}\,(a + b)} = 14$;
   (d) $y + (b - 3) = 7(4 + c)$;   (e) $\dfrac{3n}{y(5 + n)} = 7$

6. (a) 12;   (b) 6;   (c) 11;   (d) 4

7. (a) 14;   (b) 8;   (c) 8;   (d) 9

8. (a) 3;   (b) 3;   (c) 2;   (d) 1

9. (a) $d^2 + cd^2 + c^3a$;   (b) $m^3y - x^2 + px$;   (c) $\dfrac{y}{x_2} + y^2x - xy^2$;
   (d) $(c + d)^2$ or $c^2 + 2cd + d^2$;   (e) $cd + de + ef$

10. (a) Two factors of b times three factors of c; or b squared times c cubed

    (b) Three times two factors of a times f; or three a squared times f

    (c) Two factors of five times three factors of x, or five squared times x cubed

    (d) Two factors of $2y$; or the quantity $2y$ squared

11. (a) $5a + 2b$;   (b) $2ab + 2k + 3$;   (c) $a + 5b$;
    (d) $4ax^2 + by + b^2$;   (e) $xy + 4y^2$ or $y(x + 4y)$

UNIT ONE
Some Basic Concepts

Review Item	Ref Page	Example
1. The four fundamental operations in algebra are essentially the same as those of arithmetic: addition (+), subtraction (−), multiplication (X), and division (÷).	10	$6 \div 3 + \dfrac{8}{2} =$ $\dfrac{6 \times 2 + 3 \times 8}{3 \times 2}$
2. Algebra differs from arithmetic in its frequent use of letters to represent numbers.	10	Arithmetic: $2 + 3 = 5$ Algebra: $a + b = c$
3. The use of letters to represent numbers makes it possible to translate long word statements into short mathematical sentences, expressions, or statements.	10	Word statement: The difference between twice a number (n) and half that number is nine. Mathematical statement: $2n - \dfrac{n}{2} = 9$
4. A letter used to represent a number is called a *literal number* or *variable*.	11	In the equation $t + 3 = 7$, the letter t is a literal number or variable.
5. An algebraic statement that represents two things that are equal to one another is called an *equation*.	11	$8n - 3n = 5n$

Review Item	Ref Page	Example
6. The addition symbol (+) and subtraction symbol (−) are the same in algebra as in arithmetic. In arithmetic, the multiplication symbol is the "times sign," X. In algebra there are four ways of expressing the idea of multiplication. X is seldom used.	11	We could express the idea of eight times a number in any of the following ways: 8 X n, 8 · n, 8(n), or 8n
7. Like the times sign, the division symbol (÷) is seldom used in algebra. Instead, the fraction bar or, less frequently, the colon is used.	11	For $x \div y$ we would write $\frac{x}{y}$ or $x : y$ Both mean x divided by y.
8. In arithmetic, numbers being multiplied together are called *factors*. In algebra, they are referred to as *numerical factors* if they are numbers, or *literal factors* if they are letters.	11	In the expression 2xy, 2 is a numerical factor and x and y are literal factors.
9. Any factor or group of factors is the *coefficient* of the product of the remaining factors. If the factor is a number, it is called a *numerical coefficient*; if it is a letter, it is called a *literal coefficient*.	12	In the expression 2abc, 2 is the numerical coefficient of abc, and a, b, and c are the literal coefficients of 2.
10. Axioms of equality: • If equals are added to equals, the sums are equal. • If equals are subtracted from equals, the differences are equal.	13	 $4 = 6 − 2$; so, adding 2 to each side, $4 + 2 = (6 − 2) + 2$ $6 = 4 + 2$; so, subtracting 2 from each side, $6 − 2 = (4 + 2) − 2$

Review Item	Ref Page	Example
• If equals are multiplied by equals, the products are equal.		$3 + 2 = 5$; so, multiplying both sides by 2, $2 \times (3 + 2) = 2 \times 5$
• If equals are divided by equals, the quotients are equal.		$7 \times 2 = 14$; so, dividing both sides by 2, $(7 \times 2) \div 2 = 14 \div 2$
11. Division by zero is meaningless; that is, it is an undefined operation.	13	$\dfrac{5}{0}$, $\dfrac{x}{0}$, and $\dfrac{7}{a}$, where $a = 0$, are meaningless expressions.
12. When adding or multiplying, the order of the numbers may be changed without affecting the result.	14	$2 + 3 = 3 + 2$ $a + d + f = f + d + a$ $2 \cdot 3 = 3 \cdot 2$ $abc = cba$
13. When subtracting or dividing, the order of the numbers may *not* be changed.	14	$3 - 2 \neq 2 - 3$ $\dfrac{2}{3} \neq \dfrac{3}{2}$ (The symbol $\neq$ means does not equal.)
14. The sum of three or more terms or the product of three or more factors is the same regardless of how they are grouped.	15	$a + (b + c) = (a + b) + c = a + b + c$ $a(bc) = (ab)c = abc$
15. The product of an expression of two or more terms multiplied by a single factor is equal to the sum of the products of each term of the expression multiplied by the single factor.	16	$a(b + c + d) = ab + ac + ad$
16. The fundamental operations should be performed in this order:	17	In the expression $6 + 3(2) - \dfrac{4}{2}$

Review Item	Ref Page	Example
• Multiplications and divisions first, from left to right.		First multiply and divide: $6 + 6 - 2$
• Additions and subtractions next (not necessarily in order)		Then add and subtract: $6 + 6 - 2 = 10$
17. Parentheses are used: • To replace the multiplication symbol. • To group numbers. • To show that an expression should be treated as a single number.	17	$3 \times 2 = (3)(2)$ $a + (b - c)$ Double the sum of 3 and x: $2(3 + x) = 6 + 2x$
18. Parentheses can also be used to establish the order of operations when evaluating an expression.	18	In the expression $4(3 + 2)$, add the 3 and 2 in parentheses *before* multiplying by 4. Thus, $4(3 + 2) = 4 \cdot 5 = 20$
19. An *algebraic expression* is the result obtained by combining two or more numbers or letters by means of one or more of the four fundamental operations of algebra.	19	$a + b,\ 2a \div bc,\ \dfrac{x}{y},$ $\dfrac{3a + 2b}{a + b},$ and $\dfrac{3a^2}{2bc}$ are all algebraic expressions.
20. To *evaluate* (find the value of) an expression: • Substitute the given values for the letters. • Evaluate and combine terms inside parentheses.	19	Evaluate $2(x - y) + 3x - \dfrac{y}{2}$ for $x = 5$, $y = 4$. $2(5 - 4) + 3(5) - \left(\dfrac{4}{2}\right)$ $2(1) + 3(5) - \left(\dfrac{4}{2}\right)$

Review Item	Ref Page	Example
• Perform indicated multiplications and divisions. • Add and subtract as indicated.		$2 + 15 - 2$ $17 - 2 = 15$
21. A *monomial* is an expression that does not involve addition or subtraction.	20	a, $2ab$, and $\dfrac{2ab}{3bc}$ are all monomials.
22. A *multinomial* is the sum of two or more monomials. A multinomial consisting of exactly two terms is a *binomial;* one consisting of exactly three terms is a *trinomial.*	20	Binomial: $2a + 3k$ Trinomial: $a^2 + ck - 5t$ Both are multinomials.
23. Each monomial in a multinomial, together with the sign that precedes it, is called a *term* of the multinomial.	21	$2a$, $-\dfrac{3b}{2c}$, $\dfrac{a^2}{3}$, and $\dfrac{1}{2}b$ are terms of the multinomial $2a + \left(-\dfrac{3b}{2c}\right) - \dfrac{a^2}{3} + \dfrac{1}{2}b.$
24. An *exponent* is a number written to the right of and slightly above another number to indicate how many times the latter number, called the *base*, is to be taken as a factor. The product of this multiplication is called the *power*.	22	base^{exponent} = power $2^3 = 2 \cdot 2 \cdot 2 = 8$
25. In an algebraic expression, *like terms* or *similar terms* are those having the same literal coefficients (letters) and the same exponents. Algebraic expressions can be simplified by combining like terms.	22	In the expression $2a + a + 3b - b$, $2a$ and a, $3b$ and $-b$ are like terms. When simplified, $2a + a + 3b - b$ becomes $3a + 2b$.

UNIT ONE REFERENCES

1. Algebra is simply a logical extension of arithmetic. The same four funda-mental operations you learned in arithmetic are also essential in algebra: addition (+), subtraction (–), multiplication (X), and division (÷). The symbols shown are used to indicate, in mathematical shorthand, the operations to be performed. The result of addition is the sum; of subtraction, the difference or remainder; of multiplication, the product; and of division, the quotient.

2. The four operations discussed above are performed in algebra—with one major difference. In algebra *letters frequently are used to represent numbers.* Why? Because in algebra we often work with quantities without regard to their numerical values. We may need to use their numerical values eventually, but in the meantime we have to identify them in some way. So we use the letters of the alphabet.

3. How does the use of letters, numbers, and symbols make it possible to translate long word statements into brief mathematical statements? Here is an example:

Example: The sum of five times a number and two times the same number is equal to seven times the number. How can we represent this more simply?

Solution: If we let n represent the number we are talking about, we can say the same thing with this short algebraic sentence: $5n + 2n = 7n$.

Try this one: Three times a number subtracted from eight times the same number equals five times the number.

Solution: $8n - 3n = 5n$

Use letters and symbols to change these word statements into algebraic expressions:

(a) The sum of one-half x and one x equals 12. _____

(b) Twice d plus half of b added to 3 equals nine. _____

(c) Ten times a number (n) minus three times the number equals 7 more than four times the number. _____

(d) The area (A) of a triangle is equal to one-half the base (b) times the height (h). _____

_____ _____

(a) $\dfrac{x}{2} + x = 12$; (b) $2d + \dfrac{b}{2} + 3 = 9$; (c) $10n - 3n = 4n + 7$;

(d) $A = \dfrac{1}{2}bh$ or $\dfrac{bh}{2}$

4. The word *literal* means having to do with a letter (of the alphabet). In algebra, we have a special name for a letter that is used to represent a number. It is called a *literal number* or a *variable*.

5. The word statements which you translated into algebraic expressions in reference item 3 are examples of equations since, in each case, one quantity was equal to another. Bear in mind that an algebraic expression is not necessarily an equation, unless there is an equality involved. For example, $ax + by + c$ is an algebraic expression; $ax + by + c = 0$ is an algebraic expression in the form of an equation. An equation will always contain an equal sign ($=$).

6. The "times sign," $\times$, is seldom used in algebra to indicate multiplication. One reason for this is the possibility of confusing it with the letter x of the alphabet, which *does* appear frequently in algebra as a variable. As shown in the example, there are other ways of indicating multiplication. Both the dot and the use of parentheses are acceptable. Omission of the multiplication sign, as in $8n$, is preferred where either or both of the factors is a letter. Express the product of the following without using multiplication signs.

(a) $a \times b \times c$ _____   (d) $0.5 \times 40 \times t$ _____

(b) $3 \times c \times d$ _____   (e) $3 \times 4 \times dy$ _____

(c) $\dfrac{2}{5} \times 15 \times q$ _____

- - - - - - - - - - - - - -

(a) *abc*; (b) *3cd*; (c) *6q*; (d) *20t*; (e) *12dy*

7. As explained in reference item 6, the times sign ($\times$) is seldom used in algebra. The division symbol ($\div$) is used occasionally but not commonly. More frequently, the fraction bar is used to indicate division, and sometimes the colon (:). Thus, for $k^2 \div p$ we usually would write $\dfrac{k^2}{p}$ or $k^2{:}p$, both of which mean k^2 divided by p.

8. Factors, in algebra as in arithmetic, are simply numbers that are being multiplied together. If the factor is a letter, we refer to it as a *literal factor*; if it is a number, we call it a *numerical factor*. This makes it easier to talk about the various parts of an algebraic expression. For example:

Expression	Literal factors	Numerical factors
$3axy$	a, x, and y	3
$2a$	a	2
$4z\,(7\ kp)$	z, k, and p	4 and 7
$ab \cdot 9ry$	a, b, r, and y	9

Identify the literal factors in the following expressions.

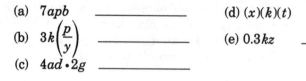

(a) $7apb$ _____   (d) $(x)(k)(t)$ _____

(b) $3k\left(\dfrac{p}{y}\right)$ _____   (e) $0.3kz$ _____

(c) $4ad \cdot 2g$ _____

- - - - - - - - - - - - - - -

(a) a, p, b;   (b) k, p, $\dfrac{1}{y}$;   (c) a, d, g;   (d) x, k, t;   (e) k, z

9. From review item 8 you know that, for example, in the expression $3xyz$, x, y, and z would be called the literal factors, and 3 would be called the numerical factor. Similarly, 3 and $(c + d)$ are factors of the expression $3(c + d)$. Now we are introducing some new terminology that will prove highly useful in the future in identifying the components of a group of factors.

Any factor or group of factors is called the *coefficient* of the product of the remaining factors. Thus, in the product $3 \cdot 5$, the number 3 is the coefficient of 5, and 5 is the coefficient of 3. In the product $4ab$, 4 is the *numerical coefficient* of ab, and ab is the *literal coefficient* of 4. If a letter does not have a coefficient written before it, the coefficient is understood to be 1. Thus, a means $1a$, x means $1x$, k means $1k$, and so on.

Check your understanding of this terminology by completing the following sentences.

(a) Numbers are represented by numerals (such as 1, 2, 3) or by

_____ when their numerical values are not given.

(b) A factor is one of two or more _____ being multiplied together.

(c) A literal factor is represented by a _____ .

(d) A numerical factor is represented by a _____ .

(e) Any factor or group of factors is the _____ of the remaining factors.

(f) Coefficients are of two kinds: _____ and _____ coefficients.

------ ---------

(a) letters; (b) numbers; (c) letters; (d) numeral; (e) coefficient;
(f) literal, numerical

10. Both arithmetic and algebra make use of the axioms of equality. An *axiom*, as you may remember, is a basic assumption that is accepted as true without proof. Axioms are considered self-evident. They are, in effect, the building blocks of mathematics. In addition to the four axioms in review item 10, another axiom you will find used frequently is this:

> *Things equal to the same thing are equal to each other.* Thus, if $a = 4$ and $b = 4$, then $a = b$. Test your understanding of these axioms by completing the following:

(a) If $5 = 7 - 2$, then $5 + 3 = $ _____

(b) If $7 = 5 + 2$, then $7 - 4 = $ _____

(c) If $\dfrac{12}{3} = 4$, then $2 \times \dfrac{12}{3} = $ _____

(d) If $4 \times 3 = 12$, then $(4 \times 3) \div 2 = $ _____

(e) If $x = 21$ and $y = 21$, then $x = $ _____

------ ---------

(a) $(7 - 2) + 3$; (b) $(5 + 2) - 4$; (c) 2×4; (d) $12 \div 2$; (e) y

11. Since, as you are now well aware, letters often are used in algebra to represent numbers, it is important to be alert to one special situation that could get you into trouble. That is the situation in which a letter stands for zero.

From arithmetic, we know that the result of adding zero to or subtracting zero from another number is the same as the original number ($x + 0 = x$; $x - 0 = x$). Nothing has changed. You probably recall also that multiplying a number by zero — or multiplying zero by a number — gives zero as a result ($x \cdot 0 = 0$). But what happens when we try to *divide* by zero?

Division by zero is an impossible operation. As shown in the example, such fractions as $\dfrac{8}{0}$ or $\dfrac{x-5}{0}$ are meaningless. This is easy to recognize when you actually see zero as the denominator (that is, the lower half of a fraction). But when the denominator contains one or more literal factors, you must be very careful that one of these letters doesn't stand for zero, or that the value assigned to the letter doesn't cause the denominator to *become* zero.

To see how this might happen, indicate in the fractions below which value of the letters in each of the denominators would result in an impossible division (that is, a zero denominator).

(a) $\dfrac{8}{c}$ _____; (b) $\dfrac{a}{2b}$ _____; (c) $\dfrac{6}{x-3}$ _____; (d) $\dfrac{8}{xy}$ _____;

(e) $\dfrac{3}{y-5}$ _____; (f) $\dfrac{2a}{3b}$ _____; (g) $\dfrac{t}{x-y}$ _____; (h) $\dfrac{0.7ax}{9.2ky}$ _____.

- - - - - - - - - - - - - -

(a) $c=0$;   (b) $b=0$;   (c) $x=3$;   (d) x or $y=0$;   (e) $y=5$;   (f) $b=0$;
(g) $y=x$;   (h) k or $y=0$.

12.   When adding, subtracting, multiplying, or dividing, the order of the numbers in an algebraic expression can sometimes be changed without affecting the result—but not in every case. If the numbers *can* be interchanged without affecting the result, the operation is said to be *commutative*. A little investigation shows that only two of the fundamental operations are commutative: addition and multiplication. This gives us the following two laws:

- *The sum of two quantities is the same whatever the order of addition.*
- *The product of two quantities is the same whatever the order of multiplication.*

Notice that these laws apply only to *pairs* of numbers, not to triples.

Indicate by the words *true* and *false* which of the following are correct examples of the commutative laws for addition and multiplication.

(a) $p+k=k+p$ _____   (e) $42+13=13+42$ _____

(b) $6+3=3+6$ _____   (f) $9b=b9$ _____

(c) $xy=yx$ _____   (g) $\dfrac{a}{5}=\dfrac{5}{a}$ _____

(d) $7-d=d-7$ _____   (h) $abc=cba$ _____

- - - - - - - - - - - - - -

(a) true;   (b) true;   (c) true;   (d) false;   (e) true;   (f) true;   (g) false;
(h) true (but for a reason we will discuss later, the commutative law for multiplication does not apply to *three* terms).

13.   Practice problems (d) and (g) in reference item 12 were false, because the commutative laws for addition and multiplication do not hold for subtraction or division. To make this clearer, suppose in problem (d) we allowed the letter d to represent the numerical value 3. We would then have

$$7-d=d-7 \text{ or } 7-3=3-7$$

which obviously is untrue.

Similarly, if in problem (g) we let the letter a represent the value 3, this would give us

$$\frac{a}{5} = \frac{5}{a} \text{ or } \frac{3}{5} = \frac{5}{3}$$

which is also obviously not true.

14. So far, in review items 12 and 13, we have considered the laws for interchanging numbers only as they relate to *pairs* of numbers. What if there are three numbers? Adding three numbers is slightly more involved. For example, if we wish to add $2 + 5 + 8$, we might first add $2 + 5 = 7$, then add $7 + 8 = 15$. But we could just as well add $5 + 8 = 13$ and then $2 + 13 = 15$. The result is the same; that is, $(2 + 5) + 8 = 2 + (5 + 8)$. To describe this property, we say that addition is *associative*. The associative law for addition states:

- *The sum of three quantities is the same regardless of the manner in which the partial sums are grouped.*

Here are some further examples:

$$a + 2 + 3 = (a + 2) + 3 = a + (2 + 3)$$
$$c + d + a = (c + d) + a = c + (d + a)$$
$$x + y + z = z + x + y = z + y + x$$

The last example combines the commutative and associative laws and illustrates the somewhat more general rule:

- *The sum of three or more numbers is the same regardless of the order in which the addition is performed.*

Similarly, if we have three *factors*, then $a \cdot b \cdot c = a(b \cdot c) = (a \cdot b)c$. This is known as the associative law for multiplication:

- *The product of three or more numbers is the same regardless of the order in which the multiplication is performed.*

Here are some examples:

$$2 \cdot 3 \cdot 4 = 2(3 \cdot 4) = (2 \cdot 3)4 = 24$$
$$c \cdot d \cdot f = c(d \cdot f) = (c \cdot d)f = cdf \text{ or, simply,}$$
$$cdf = c(df) = (cd)f$$

Based on what we have covered so far about the commutative and associative laws, determine whether the following statements are *true* or *false*.

(a) $7xy = yx7$ _____

(b) $k + r + t = r + k + t$ _____

(c) $pt(z) = (p)zt$ _____

(d) $a + b - c = b + a - c$ _____

(e) $2x \div y = 2y \div x$ _____

(f) $k + m - n = n + k - m$ _____

(a) true; (b) true; (c) true; (d) true (because the position of the number being subtracted was not changed); (e) false; (f) false (because the position of the number being subtracted was changed)

Once more, then:

- When *adding,* you *may* change the order of the numbers.
- When *subtracting,* you *may not* change the order of the numbers.
- When *multiplying,* you *may* change the order of the numbers.
- When *dividing,* you *may not* change the order of the numbers.

15. In addition to the commutative and associative laws, there is a third law known as the *distributive* law for multiplication. This law states:

- *The product of an expression of two or more terms by a single factor is equal to the sum of the products of each term of the expression by the single factor.*

In simpler mathematical language this law says that

$$a(b + c) = ab + ac$$

or, using numbers instead of letters,

$$2(3 + 4) = 2 \cdot 3 + 2 \cdot 4$$

Before considering further applications of the distributive law, you need to recall that if a number, such as a, is multiplied by itself, we write $a \cdot a = a^2$. Similarly, $a \cdot a \cdot a = a^3$. The exponent (that is, the number written to the right and a little above the number being multiplied by itself, in this case the letter a) indicates the number of times the quantity a is used as a factor. (See reference item 24 below.)

With this in mind, here are a few more examples of the distributive law:

$$a(a + b) = a^2 + ab$$
$$2b(ab + bc) = 2ab^2 + 2b^2c$$

For more than two terms we use the *extended distributive* law:

$$2a(2a + 3b - 4ad) = 4a^2 + 6ab - 8a^2d$$
$$3ab(a^2 - 2ad + b) = 3a^3b - 6a^2bd + 3ab^2$$

Apply the extended distributive law to the following multiplication problems.

(a) $b(c + d + e) =$ _____

(b) $bc(c + d - 2) =$ _____

(c) $3x(2 - xy + z) =$ _____

(d) $4ab(2ab + 3ac - cd) =$ _____

- - - - - - - - - - - - -

(a) $bc + bd + be$;   (b) $bc^2 + bcd - 2bc$;   (c) $6x - 3x^2y + 3xz$;
(d) $8a^2b^2 + 12a^2bc - 4abcd$

16.   When more than one algebraic operation is indicated in an expression, the various operations should be performed in the right order for best results. The rule is:

- Do multiplications and divisions first, in order from left to right.
- Do additions and subtractions second.

For example, in the expression $6 + 3(2) - \dfrac{4}{2}$ performing the multiplication gives us $6 + 6 - \dfrac{4}{2}$. Next, performing the division we get $6 + 6 - 2$. Finally, adding and subtracting gives us the answer: 10.

At the moment, however, the answer is not nearly as important as the *procedure* you follow. Study the expression below and indicate the correct sequence of operations.

$$\frac{3 \cdot 4}{6} + 2(7) - \frac{9}{3} =$$

	Operation	Term(s)
1.	_____	_____
2.	_____	_____
3.	_____	_____

- - - - - - - - - - - - -

1. multiplications $3 \cdot 4$ and $2(7)$
2. divisions   12 by 6 and 9 by 3
3. addition and subtraction   All terms (=13)

17.   In working with the associative law for addition, we used parentheses to group numbers: $a + (b - c)$. However, we also use parentheses to indicate multiplication (as covered in review item 6) and to show that an expression should be treated as a single number. Thus, if we wish to double the sum of 3 and x, we write $2(3 + x)$. From working with the distributive law, you know that this tells us that we must multiply both 3 and x by 2 to get the correct answer. Or if we wished to multiply the difference, 9 minus y, by 3, we would write this as $(9 - y)3$ or $3(9 - y)$; either is correct.

Below are some further examples of the use of parentheses to express word statements algebraically:

1.	The sum of k and twice p.	$k + 2p$
2.	Twice the sum of s and r.	$2(s + r)$
3.	Twice the sum of a plus b equals 9.	$2(a + b) = 9$
4.	a divided by the sum of a and b plus twice xy equals 7.	$\dfrac{a}{a + b} + 2xy = 7$

Now here are a few for you to practice on. Use parentheses to express the following relationships.

(a) Twice the sum of $c + d$ equals 7. _____

(b) b divided by the sum of b and a, plus twice qp equals 9. _____

(c) Two added to one-third the quantity of y minus z equals $2z$. _____

(d) a plus half the quantity of y minus 2 equals 13. _____

(e) Three times a number n divided by y times the sum of 1 and the number is equal to 7. _____

- - - - - - - - - - - - - -

(a) $2(c + d) = 7$;   (b) $\dfrac{b}{b + a} + 2qp = 9$;   (c) $2 + \dfrac{1}{3}(y - z) = 2z$;

(d) $a + \dfrac{1}{2}(y - 2) = 13$; (e) $\dfrac{3n}{y(1 + n)} = 7$

18. Here are a few problems to help you practice evaluating expressions containing parentheses. Remember the order of operations: Multiplication and division first, addition and subtraction last. Terms enclosed in parentheses should be combined wherever possible.

Find the value of each of the following expressions.

(a) $2(3 + 4) - 2 \cdot 3 =$ _____

(b) $6 - \dfrac{1}{2}(4 - 2) =$ _____

(c) $3 - \dfrac{1}{3}(4 + 2) =$ _____

(d) $12 - 5(4 - 2) =$ _____

(e) $2(3 + 2) - 7 + \dfrac{9}{3} = -$_____

(f) $\dfrac{6 + 4}{2} - \dfrac{6}{3} + 2 \cdot 4 =$ _____

(g) $7 - \dfrac{12}{4} + 2(3 + 1) =$ _____

(h) $\dfrac{9 + 3}{4 - 1} - 3 + (6 \div 2) =$ _____

Identify the following as either monomials, binomials, trinomials, or multinomials.

(a) $ab + jk + 4$ _____ (d) $\dfrac{pk}{7} + r^2 - 3$ _____

(b) $\dfrac{2xy}{3k}$ _____ (e) $3c^2 + 12$ _____

(c) $7 - ax + \dfrac{y}{3} + z^2$ _____ (f) $2(a + b) - 8$ _____

- - - - - - - - - - - - - -

(a) trinomial;   (b) monomial;   (c) multinomial;   (d) trinomial;   (e) binomial;
(f) binomial

23.   Up to this point we have used the word "term" in a rather general sense. Now we can be more explicit, since we have previously defined the word multinomial.

The only part of the definition of a *term* given in review item 23 that may seem new to you is that the *sign* is a part of the term—and an important part.

In Unit 2, we will discuss the fact that (+) and (–) symbols can be used either as signs of operation (that is, telling us to add or subtract quantities) or as indications that the quantities themselves are positive or negative. Although we will not go into this in any detail now, the example shown illustrates this idea. Here again is the multinomial used:

$$2a + \left(-\frac{3b}{2c} \right) - \frac{a^2}{3} + \frac{1}{2}b$$

The second term is negative, as shown by the minus symbol. But since we wish to add it to the first term, the sign of operation is plus (+). The parentheses are used simply to separate the two signs. Since, by the rules of algebraic addition, adding a minus quantity is the same as subtracting a positive quantity, when writing this multinomial we would normally omit the parentheses and change the sign of operation to indicate subtraction. Thus, the multinomial would appear as

$$2a - \frac{3b}{2c} - \frac{a^2}{3} + \frac{1}{2}b$$

How many terms are in each of the following expressions?

(a) $4z + \dfrac{1}{3}bx - 3(k - z)$ _____   (c) $c(d) + b^2c - \dfrac{dx}{3}$ _____

(b) $3(c + d) - \dfrac{y}{x} + 2z$ _____   (d) $bc + cd - de(y + x)$ _____

- - - - - - - - - - - - - -

(a) 3;   (b) 3;   (c) 3;   (d) 3

24. We touched briefly on the subject of exponents in review item 15. Now it is time to take a closer look at them. The repeated multiplying of a factor by itself is an important concept since it occurs regularly in algebraic expressions. For example, if we wish to multiply $2 \cdot 2 \cdot 2$, we may express this in shorter form by writing 2^3. This is read as "two cubed" or "two to the third power." Similarly, the expression x^2y^3 would be read "x squared times y cubed." What this last example means is "two factors of x times three factors of y." (*Note:* If a numeral or letter has *no* exponent written at its upper right, the exponent is understood to be 1. Thus, y means y^1 and 4 means 4^1.)

Write the following expressions using exponents where appropriate:

(a) $cc + acc + bbbc$ _____

(b) $mmmy - xx + mx$ _____

(c) $\dfrac{y}{mm} + xyy - m(my)$ _____

(d) $ab + bc + cd$ _____

What do the following expressions mean?

(e) a^2b^3 _____

(f) $7d^2e$ _____

(g) 3^2x^3 _____

(h) $(4y)^2$ _____

- - - - - - - - - - - - - - -

(a) $c^2 + ac^2 + b^3c$;   (b) $m^3y - x^2 + mx$;   (c) $\dfrac{y}{m^2} + xy^2 - m^2y$;   (d) $ab + bc$ $+ cd$ (no repeated factors);   (e) two factors of a times three factors of b (or a squared times b cubed);   (f) 7 times two factors of d times e;   (g) two factors of 3 times three factors of x;   (h) two factors of $4y$ (or the expression $4y$ squared)

25. The process of simplifying an algebraic expression is merely a matter of combining *like* (or *similar*) *terms*. Therefore, it is important to be able to recognize similar terms. As indicated in review item 25, terms are similar if they have the same literal coefficients *and* these coefficients have the same exponents. Thus, in the expression $3a + 4a$, the two terms can be combined to become $7a$ because the literal coefficients are the same and have the same exponents (a to the first power in each case). In the expression $2a + 3x^2 + 3a + x^2$, the like terms can be combined to produce the simplified expression $5a + 4x^2$. Similarly,

$$6a + 2a^2 - 3a - a^2 = 3a + a^2$$

and

$$2ab + cd^2 + cd - ab = ab + cd + cd^2$$

Simplify the following expressions where possible:

(a) $3a + 4a - 7 = $ _____

(b) $4a - 3b + 2a + 6b = $ _____

(c) $2k + 4k - 8c - 8$ _____

(d) $2x^2 - 3y^2 + 4x^2 - 13 = $ _____

(e) $x^2 + y^2 + 7x^2 - 2y^2 = $ _____

(f) $2xy - 3ak + 3xy + 4ak = $ _____

(g) $3y^2 + x - y^3 + 2xy = $ _____

(h) $2x + x^2 + 3x - 3x^3 = $ _____

- - - - - - - - - - - - - - -

(a) $7a - 7$; (b) $6a + 3b$; (c) $6k - 8c - 8$; (d) $6x^2 - 3y^2 - 13$; (e) $8x^2 - y^2$; (f) $5xy + ak$; (g) $3y^2 + x - y^3 + 2xy$ (note that no combining of terms was possible); (h) $5x + x^2 - 3x^3$

UNIT TWO

The Number System

Review Item	Ref Page	Example
1. The *natural* or *counting numbers* are the whole numbers greater than zero. Together with *fractions,* they are the numbers we use in arithmetic.	31	Whole numbers: 1, 2, 3, 4, . . . Fractions: $\dfrac{1}{4}, \dfrac{1}{2}, \dfrac{2}{3}, \dfrac{3}{2}, \ldots$
2. *Negative numbers* are numbers whose values are less than zero. They are the opposite of *positive numbers.*	31	$-1, -4, -7, -11, \ldots$ $-\dfrac{1}{2}, -\dfrac{3}{4}, -\dfrac{7}{8}, \ldots$
3. The set of all positive and negative whole numbers, together with zero, is known as the set of *integers.*	31	$0; +1, -1; +2, -2; +3, -3; \ldots$
4. The set of all integers and positive and negative fractions is known as the set of *rational numbers.* A rational number is one that can be expressed as the quotient, or ratio, of two integers.	32	All the examples given for review items 1 to 3 are rational numbers.
5. *Irrational numbers* are numbers that *cannot* be expressed as the ratio of two integers.	32	$\sqrt{2}$ (the square root of 2) π (pi; the ratio of the circumference of a circle to its diameter)

Review Item	Ref Page	Example
6. *Signed numbers* include both positive and negative numbers. They are called signed numbers because they require signs to tell them apart.	33	The positive or plus numbers (above zero) and the negative or minus numbers (below zero) on the temperature scale of a thermometer. +60 +50 +40 +30 +20 +10 0 −10 −20 −30
7. A positive number is indicated by a plus (+) sign or by no sign at all; a negative number is indicated by a minus (−) sign.	33	+2, −3, +7, − 4, 5, 8, . . .
8. The *absolute value* of a signed number is the number obtained by disregarding the sign. Absolute value represents magnitude (size) but not direction (positive or negative). Absolute value is generally written as $\lvert n \rvert$. $\lvert x \rvert = x ; \lvert -x \rvert = x.$	34	The absolute values of +2, − 6, and +3 are 2, 6, and 3. $\lvert 3 \rvert = 3$ $\lvert 0 \rvert = 0$ $\lvert -3 \rvert = 3$
9. Plus and minus symbols are used in two ways in algebra: first, as in arithmetic, to indicate what *operation* to perform on numbers (i.e., whether to add or subtract them); second, to indicate the *quality* of a number (i.e., whether it is positive or negative). In the first use they are called *signs of operation;* in their second use, *signs of quality* or *signs of condition.*	35	Signs of operation: $2 + 4 = 6$ $4 - 2 = 2$ Signs of quality or of condition: − 4, +2, −3, +1

Review Item	Ref Page	Example
10. To avoid confusion between signs of operation and signs of quality, it is the practice to use a raised minus sign to indicate a negative number. Parentheses are also used to help avoid confusion. Thus negative 3 is written (⁻3).	35	The expression +3 + (⁻4) means we are to add negative 4 to positive 3.
11. To help visualize the process of adding and subtracting positive and negative numbers, we use vertical and horizontal scales.	35	Signed numbers / Absolute value +5 — 5 +4 — 4 +3 — 3 +2 — 2 +1 — 1 0 — 0 ⁻1 — 1 ⁻2 — 2 ⁻3 — 3 ⁻4 — 4 ⁻5 — 5 ⁻6 — 6 Signed numbers ⁻6 ⁻5 ⁻4 ⁻3 ⁻2 ⁻1 0 +1 +2 +3 +4 +5 6 5 4 3 2 1 0 1 2 3 4 5 Absolute value
12. When we speak of one number being greater or less than another, we are referring to the relative locations of the two numbers on the number scale. The *value* of a number is its position on the number scale. Hence value includes both distance *and* direction from zero.	36	greater → −4 −3 −2 −1 0 +1 +2 +3 +4 +2 is greater than ⁻4 because it is further to the right or higher than ⁻4 on the number scale.

Review Item	Ref Page	Example
13. To add (combine) two numbers with *like* signs, add their absolute values and prefix the common sign.	37	$(+2) + (3) = +5$ $(^-4) + (^-6) = ^-10$
14. To add two numbers with *opposite* signs, subtract the smaller absolute value from the larger and prefix the sign of the number whose absolute value is larger.	37	$(+4) + (^-6) = ^-2$ $\|^-6\| - \|+4\| = ^-2$
15. When writing additions horizontally, as we usually do in algebra, we may omit the signs of operation and parentheses and use only the signs of quality. Also, if the first signed number is positive, its plus sign may be omitted.	37	For $(+2) + (^-3) = ^-1$, write $2 - 3 = ^-1$. For $(+3) + (+8) + (^-9)$, write $3 +8 -9$.

Review Item	Ref Page	Example
16. Algebraic subtraction can best be visualized by considering it as finding the *directed distance* from one position to another on the number scale. This involves both distance and direction. The distance (the number of scale units) between two positions gives the absolute value of the answer; the direction in which we count determines the sign of the answer.	38	Find the directed distance from +3 to ⁻3. Distance: 6 Direction: Downward (negative)
17. To subtract using the number scale, count *from the subtrahend* (the number being subtracted) *to the minuend* (the number from which it is being subtracted).	38	To subtract +2 from +4, count from +2 to +4, two spaces upward, giving an answer of +2. To subtract ⁻2 from +4, count from ⁻2 to +4, six spaces upward, resulting in an answer of +6.

Review Item	Ref Page	Example
		To subtract +4 from ⁻2, count from +4 to ⁻2, six spaces downward, giving an answer of ⁻6. +4 +3 +2 +1 ⁻6 0 ⁻1 ⁻2 To subtract ⁻2 from ⁻4, count from ⁻2 to ⁻4, two spaces downward, giving an answer of ⁻2. +1 0 ⁻1 ⁻2 ⁻3 ⁻2 ⁻4
18. A simple rule for algebraic subtraction not requiring the use of a number scale is: *To subtract a signed number, add its opposite.*	40	$(+6) - (+2) = (+6) + (⁻2)$ or $6 - 2 = 4$ $(+6) - (⁻2) = (+6) + (+2)$ or $6 + 2 = 8$
19. In multiplying two signed numbers, the rule is: *If the signs are the same, the product is positive; if the signs are different, the product is negative.*	41	$(2)(3) = 6$ $(⁻2)(3) = ⁻6$ $(⁻2)(⁻3) = 6$ $(2)(⁻3) = ⁻6$

Review Item	Ref Page	Example
20. Regardless of the number of factors, the product of more than two numbers is always *negative* if there are an odd number of negative factors, and *positive* if there are an even number of negative factors.	42	Since the expression $(2)(-3)(-2)$ contains an even number of negative factors, the product is $+12$. Since the expression $(-2)(-3)(-2)$ contains an odd number of negative factors, the product is -12.
21. *Odd powers* of negative numbers are *negative; even powers* of negative numbers are *positive.*	42	$(-2)^3 = -8$ $(-2)^4 = +16$
22. In dividing two signed numbers, if the signs are *alike,* the quotient will be *positive* ; if the signs are *different,* the quotient will be *negative.*	43	$-\dfrac{4}{-2}$ or $\dfrac{+4}{+2} = +2$ $\dfrac{-4}{+2}$ or $\dfrac{+4}{-2} = -2$
23. When performing combined multiplication and division of signed numbers, first multiply the factors in the numerator; multiply the factors in the denominator; then divide.	43	$\dfrac{(-3)(-6)}{(+2)(-3)} = \dfrac{+18}{-6} = -3$ $\dfrac{(-1)(-7)(+6)}{(+2)(-1)(+3)} = \dfrac{+42}{-6} = -7$
24. To evaluate expressions containing signed numbers, first substitute the values given for the letters, enclosing them in parentheses; then perform the indicated operations in the correct order.	44	Evaluate $2xy - \dfrac{6}{y}$ for $x = -3$, $y = 4$. $2(-3)(4) - \left(\dfrac{6}{-3}\right)$ $-24 - (-2)$ $-24 + 2 = -22$

UNIT TWO REFERENCES

1.   The numbers you first learned to count with are called *natural numbers* or *counting numbers*. They are the whole numbers greater than zero.

Later you learned how to add, subtract, multiply, and divide these numbers. You soon found, however, that while some divisions (such as 8 ÷ 4 or 9 ÷ 3) resulted in a whole number as a quotient, others (such as 5 ÷ 2 or 7 ÷ 3) did not. A new class of numbers, known as *fractions,* had to be introduced to give meaning to the results of such divisions.

Another name for natural numbers is _____ numbers.

_ _ _ _ _ _ _ _ _ _ _ _ _ _ _

counting

2.   The number system of arithmetic consists of whole numbers, zero, and fractions. But there is a handicap to this system of numbers. Within this system, we cannot, for example, subtract a large number from a smaller number. To overcome this problem, mathematicians invented *negative numbers—* numbers that are less than zero. For every positive number, there exists a number that is the negative (opposite) of the *positive number.*

The opposite of a positive number is known as a _____ .

_ _ _ _ _ _ _ _ _ _ _ _ _ _ _

negative number

3.   We began in arithmetic with the set of natural numbers, along with zero. When we include the corresponding negative numbers, we have what is known as the set of *integers,* as shown below.

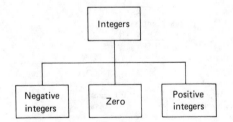

Another name for the set of positive and negative whole numbers, together with zero, is _____ .

_ _ _ _ _ _ _ _ _ _ _ _ _ _ _

integers

4. The entire collection of integers and positive and negative fractions is known as the set of *rational numbers,* as shown below.

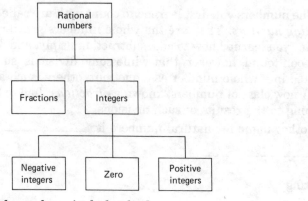

The set of rational numbers includes both _____ and _____
_____ .

_____ _____

integers, fractions

5. The counterpart of rational numbers is *irrational numbers*—numbers that cannot be expressed as the ratio of two integers. Since you need to know nothing about irrational numbers at the moment, other than the fact that they exist, we will mention only two examples of such numbers, shown also in your review item example 5: π (pi; the ratio of the circumference of a circle to its diameter) and $\sqrt{2}$ (the square root of 2). Neither can be expressed as an ordinary fraction.

The rational numbers together with the irrational numbers constitute the entire family of *real numbers,* as shown below.

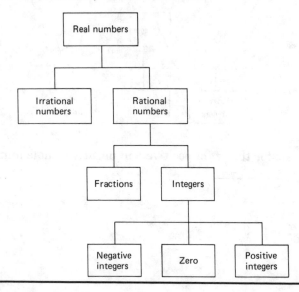

An irrational number is one that cannot be expressed as the _____ of two integers.

- - - - - - - - - - - - - - - -

ratio

6. The diagram above represents all the elements of the real number system and, therefore, all the numbers with which we will be concerned. In order to be able to refer to positive and negative numbers properly, we call them *signed numbers.* Although zero is neither positive nor negative, we include it with the signed numbers.

You already are familiar with several kinds of signed numbers. A thermometer, for example, has a scale containing both positive numbers (numbers above zero) and negative numbers (numbers below zero). Just as +15° represents 15° above zero, −15° represents 15° below zero. Plus and minus numbers on a temperature scale, therefore, constitute a set of opposites. The concept of a number scale or number line similar to the scale on a thermometer is useful in discussing signed numbers.

```
+100 —
 +90 —
 +80 —
 +70 —
 +60 —
 +50 —
 +40 —
 +30 —
 +20 —
 +10 —
   0 —
 −10 —
 −20 —
 −30 —
 −40 —
 −50 —
```

What does a thermometer reading of +45° mean? _____

What does a thermometer reading of −12° mean? _____

- - - - - - - - - - - - - - -

45° above zero; 12° below zero.

7. This is an important point to remember when working with signed numbers: *A negative number must be indicated by the use of a minus sign; a positive number may be indicated by a plus sign or by no sign at all.*

To develop further this idea of opposite values, answer the following questions.

(a) If +25 means a gain of $25, what does −25 mean? _____

(b) If +10 means 10 steps forward, what does −10 mean? _____

(c) If −18 inches means 18 inches below the water line, what does +18 mean? _____

(d) If +3 means a positive force of 3g, what does −3 mean? _____

- - - - - - - - - - - - - - -

(a) a loss of $25   (b) 10 steps back;   (c) 18 inches above the water line;
(d) a negative force of $3g$

8.

$$-6\ -5\ -4\ -3\ -2\ -1\ \ 0\ +1\ +2\ +3\ +4\ +5\ +6\ +7\ +8$$

The figure above is called a *number line* or *number scale*. The representation of real numbers as points on a line is an important concept in establishing the connection between arithmetic and geometry. Every point corresponds to one and only one real number, and every real number corresponds to one and only one point.

In order to be able to write simple and meaningful rules for working with signed numbers, we need to give a name to the distance between a number and zero on the number line. We call this distance the *absolute value*. When using the number scale, we consider the absolute value to mean the distance between a number and zero without regard to the direction from zero.

By looking at the number scale, we can see that $+3$ and -3 must have the same absolute value, since they are each three units from zero. Hence, both have an absolute value of 3. Similarly, $+5$ and -5 are each five units from zero, and therefore the absolute value of each is 5. Notice also that the absolute value of -5 is *greater* than the absolute value of $+3$, even though the number $+3$ is larger than the number -5.

The concept of absolute value occurs so frequently that a special symbol is used to represent the absolute value of a number. We use $|n|$ to mean the absolute value of the number n. Thus, $|4| = 4$ is read "the absolute value of $+4$ is 4," and $|-6| = 6$ is read "the absolute value of -6 is 6."

Apply these concepts to the problems below.

(a)   The absolute values of 25 and -8 are ____ and ____.

(b)   The absolute values of 10 and -10 are ____ and ____.

(c) $|12|$ equals ____ and $|-30|$ equals ____.

Write the absolute value of each pair of numbers and the difference of their absolute values.

	Absolute Values	*Difference*
(d) +21 and –16	_____	_____
(e) –17 and –9	_____	_____
(f) –35 and +25	_____	_____
(g) +4½ and 3	_____	_____

- - - - - - - - - - - - - -

(a) 25 and 8;   (b) 10 and 10;   (c) 12 and 3;   (d) 21 and 16, 5;   (e) 17 and 19, 8; (f) 35 and 25, 10;   (g) 4½ and 3, 1½.

9. In arithmetic, the plus (+) sign always indicates addition, and the minus sign (–) subtraction. They are called *signs of operation* because they tell us what mathematical operation to perform on the numbers. In algebra, as you know by now, plus and minus signs may also be used to indicate that numbers are positive or negative. When the signs are used in this way they are called *signs of quality* or *signs of condition.*

10. Since, as we discussed in review item 9, the plus and minus signs can be used in algebra to indicate an operation to be performed *or* the quality of a number, to avoid confusion we will (at least for the time being) adopt the practice of writing ⁻3 when we mean negative 3. (Notice that the position of the minus sign is higher than you are accustomed to finding it.) Parentheses are also used to help avoid confusion. Thus, the expression +3 + (⁻4) means we are to add positive three to negative four.
Write:

(a) Negative 5 added to negative 8 _____

(b) Positive 4 added to negative 6 _____

(c) Negative 3 added to positive 7 _____

(d) Positive 9 subtracted from negative 2 _____

(e) Negative 5 subtracted from negative 8 _____

- - - - - - - - - - - - - - -

(a) ⁻5 + (⁻8);   (b) 4 + (⁻6);   (c) ⁻3 + 7;   (d) ⁻2 – 9;
(e) ⁻8 – (⁻5)

11. So far, we have not talked about how to work with signed numbers. To explain how to add and subtract positive and negative numbers, we make use of number scales. These are of two kinds: vertical and horizontal. The *vertical number scale* shown is similar to the temperature scale shown in review item 6. Note again that positive and negative numbers that are the same distance from zero have the same absolute value.

Signed Numbers	Absolute value
+10	10
+9	9
+8	8
+7	7
+6	6
+5	5
+4	4
+3	3
+2	2
+1	1
0	0
⁻1	1
⁻2	2
⁻3	3
⁻4	4
⁻5	5
⁻6	6
⁻7	7
⁻8	8
⁻9	9
⁻10	10

Similarly, in the *horizontal number scale* shown below, the positive and negative numbers appear on opposite sides of zero, which is also known as the *origin*. The absolute values of the positive numbers increase to the right, and the absolute values of the negative numbers increase to the left.

Signed numbers

⁻10 ⁻9 ⁻8 ⁻7 ⁻6 ⁻5 ⁻4 ⁻3 ⁻2 ⁻1 0 +1 +2 +3 +4 +5 +6 +7 +8 +9 +10

10 9 8 7 6 5 4 3 2 1 0 1 2 3 4 5 6 7 8 9 10

Absolute value

On both scales, the positive numbers represent equally spaced distances from the origin in one direction, and the negative numbers similar distances from the origin in the opposite direction.

12. Number scales help us understand the meaning of signed numbers. They also provide a convenient way of visualizing which of two signed numbers is the greater. But what do we mean by "greater"?

> *On a vertical number scale greater means higher; on a horizontal number scale greater means further to the right.*

The positive direction of the vertical number scale is upward, and the positive direction of the horizontal number scale is to the right. When we compare two numbers, we determine the value of one number in comparison with the other by taking into account the absolute values and the signs of the numbers.

It is not unusual to find that the absolute value of one number is less than that of another although its relative value is greater. For example, +10° is greater (that is, higher on the temperature scale) than –30°; yet its absolute value, 10, is numerically less than 30. In fact, *any* positive number is greater than any negative number, although it may or may not have a greater absolute value.

⁻10 ⁻9 ⁻8 ⁻7 ⁻6 ⁻5 ⁻4 ⁻3 ⁻2 ⁻1 0 +1 +2 +3 +4 +5 +6 +7 +8

You may find the horizontal number line above helpful in answering the following questions.

(a) Which is greater, +2 or ⁻9? _____

(b) Is zero greater than or less than ⁻ 7? _____

(c) How many units apart are 8 and ⁻4? _____

(d) Is ⁻3 greater than or less than ⁻4? _____

(e) If you start at +3 and count five spaces to the left, what number will you stop at? _____

(f) How would you express in mathematical symbols the operation you performed in problem (e)? _____

- - - - - - - - - - - - - -

(a) +2; (b) greater; (c) 12 units; (d) greater; (e) ⁻2; (f) +3 − 5 = ⁻2

13. In algebra, when we say "add," we really mean *combine*. Thus, when we speak of adding two numbers, we mean that we are combining them to obtain a single number that represents the total or combination of the two. Here is the rule again:

> *To add two numbers with like signs, add their absolute values and prefix their common sign.*

Use this rule to perform the following additions.

(a) (+8) + (+4) = ____   (c) (+3) + (+7) = _____

(b) (⁻4) + (⁻13) = ____   (d) (⁻1) + (⁻9) = _____

- - - - - - - - - - - - - -

(a) +12; (b) ⁻ 17; (c) +10; (d) ⁻ 10

14.   *To add two numbers with opposite signs, subtract the smaller absolute value from the larger and prefix the sign of the number whose absolute value is larger.*

Above is the rule once more, for your reference, and herewith another example.

$$(⁻7) + (+8) = |8| - |⁻7| = 8 - 7 = 1$$

Practice using this rule in the following problems.

(a) (⁻3) + (+10) = ____   (c) (⁻6) + (+1) = _____

(b) (+10) + (⁻8) = ____   (d) (+4a) + (⁻2a) = _____

- - - - - - - - - - - - - -

(a) +7; (b) +2; (c) ⁻5; (d) +2a

15. Another example of the principle described in review item 15 would be:

For (+3) + (+8) + (⁻9), write 3 + 8 −9.

Notice that, first, we eliminated all parentheses. Then we dropped the signs of operation (the two plus signs between the parentheses), putting in their place the signs of quality appearing before each number. Finally, since the sign in front of the first number was a plus sign, we dropped it.

Rewrite the following, using a minimum of signs and parentheses.

(a) $(+8) + (^-4) + (+2) =$ _____

(b) $(+2) + (+3) + (^- 8) =$ _____

(c) $(+8) + (^- 6) + (^-4) =$ _____

(d) $(+2) + (^-4) + (+3) + (^- 6) =$ _____

(e) $(^- 2) + (^-4) + (+3) =$ _____

Express the following with a minimum number of symbols, and then add them.

(f) $(+3) + (+7) + (^-1) + (^-5) =$ _____ = _____

(g) $(^- 13) + (^- 2) + (+30) + (^-6) =$ _____ = _____

(h) $(+3) + (^- 3) + (+11) + (+4) =$ _____ = _____

(i) $(^- 1) + (^- 2) + (^- 3) - (+4) =$ _____ = _____

(a) $8 - 4 + 2$;   (b) $2 + 3 - 8$;   (c) $8 - 6 - 4$;   (d) $2 - 4 + 3 - 6$;   (e) $-2 - 4 + 3$;
(f) $3 + 7 - 1 - 5 = +4$;   (g) $-13 - 2 + 30 - 6 = +9$;   (h) $3 - 3 + 11 + 4 = +15$;
(i) $-1 - 2 - 3 - 4 = -10$

16.   So far, we have talked only about adding signed numbers. Now we need to consider the procedures for *subtracting* signed numbers. Because we worked only with numbers greater than zero in arithmetic, we could not subtract a larger number from a smaller one. In algebra we can. In fact, negative numbers were invented in order to make this kind of subtraction possible.

17.   In problem (e) of reference item 12, you subtracted 5 from 3 using the horizontal number scale. Now let us subtract 3 from 5 using the vertical number scale at the right. To do this, count the number of spaces between +3, the *subtrahend* (the number being subtracted), and +5, the *minuend* (the number from which 3 is being subtracted). The scale distance is 2 and the direction (from 3 to 5) is upward (positive). Hence the difference is +2. The complete problem would be written horizontally this way:

minuend		subtrahend		difference
5	−	(+3)	=	+2

Subtract 5 from 3 and explain algebraically how you arrived at your answer.

- - - - - - - - - - - - - -

To subtract 5 from 3, count the number of spaces between +5 (the subtrahend) and +3 (the minuend). The distance is 2 and the direction is downward (negative). The answer is ⁻2.

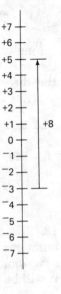

Now let us consider how to subtract a negative number from a positive one. You have seen one example of this. Here is another.

To subtract ⁻3 from +5, count from ⁻3 to +5. The distance is 8 and the direction is upward (positive). The difference, therefore, is +8, as shown on the vertical scale below.

Remember: Always count *from* the subtrahend *to* the minuend. This determines the direction in which you are counting and therefore the sign of the answer.

Write out below, in a horizontal line, the algebraic solution to the subtraction performed above. (Refer to the example of this given above if you need some help.) _____

- - - - - - - - - - - - - -

$5 - (⁻3) = +8$

Having learned how to subtract a negative number from a positive one, let us reverse the procedure and subtract a positive number from a negative one, using the same numbers as before. Thus, to subtract +3 from ⁻5, count from +3 to ⁻5. The distance is 8 and the direction is downward; hence, the difference is ⁻8, as shown on the number scale below.

Write this operation in the form $a - b = c$. _____

---------- ----------

$-5 - (+3) = -8$

Finally, let us subtract a negative number from another negative one. We will subtract -3 from -5. The distance is 2 units and the direction downward (negative); hence, the difference is -2. Write this problem in the form $a - b = c$. _____

---------- ----------

$-5 - (-3) = -2$

```
+4
+3
+2
+1
 0
-1
-2
-3
-4   -2
-5
-6
```

18. The preceding examples illustrate the process of subtracting various combinations of signed numbers. Before working some problems, let us restate, in somewhat more formal terms, the rule for subtracting signed numbers so that it will be handy to refer to:

> *To subtract a number, add (using the correct rule for addition) the negative (opposite) of that number. In symbols: $a - b = a + (-b)$.*

To see how this rule works, let us summarize what we have covered.

Examples above based on use of number scale	Same problems worked out by rule for subtraction
$+5 - (+3) = +2$	$+5 - (+3) = +5 + (-3) = 2$
$+5 - (-3) = +8$	$+5 - (-3) = +5 + (+3) = 8$
$-5 - (+3) = -8$	$-5 - (+3) = -5 + (-3) = -8$
$-5 - (-3) = -2$	$-5 - (-3) = -5 + (+3) = -2$

Use the rule of subtraction to solve these problems.

Example: $(-3) - (+5) = -3 + (-5) = -8$

   (a) $(+7) - (^-4) =$ _____ = ____

   (b) $(^-8) - (^-2) =$ _____ = ____

   (c) $(7) - (3) =$ _____ = ____

   (d) $1 - 4 =$ _____ = ____

   (e) $1 - (^-4) =$ _____ = ____

Combine terms.

   (f) $(+4) - (^-2) =$ ____   (h) $(+18) - (+3) - (+9) =$ ____

   (g) $(^-7) - (^-9) - (+4) =$ ____   (i) $8 - 4 + 2 - 12 =$ ____

------ ----------

(a) $+7 + (+4) = 11$; (b) $^-8 + (^+2) = ^-6$; (c) $7 + (^-3) = 4$; (d) $1 + (^-4) = ^-3$;
(e) $1 + (+4) = 5$; (f) $+6$; (g) $^-2$; (h) $+6$; (i) $^-6$

19.   Below are some problems that will give you practice in applying the rule for multiplying signed numbers. But first, note this change in the procedure we have been following:

> *Since you now understand the difference between the minus symbol (–) as a sign of quality and as a sign of operation, we will discontinue the practice of raising it above the usual position to indicate a negative number.*

We will, however, continue to use parentheses to show that the minus sign is associated with a number. Thus, instead of $(^-3)(^-4) = 12$, we will write $(-3)(-4) = 12$. Now to the problems.

   (a) $(+9)(+4) =$ ____   (g) $(+2)(8)$ _____

   (b) $(-7)(-3) =$ _____   (h) $(3)(+12) =$ ____

   (c) $(5)(6) =$ ____   (i) $7(5) =$ ____

   (d) $(5)(-5) =$ ____   (j) $(-1)(+1) =$ ____

   (e) $(-7)(2) =$ ____   (k) $6(-2) =$ ____

   (f) $(0.5)(-8) =$ ____   (l) $-4 \cdot 4 =$ ____

------ ----------

(a) 36;   (b) 21;   (c) 30;   (d) –25;   (e) –14;   (f) –4;   (g) 16;   (h) 36;   (i) 35;
(j) –1;   (k) –12;   (l) –16.

20.   The inclusion and omission of plus signs and parentheses in the above problems was intended to help you recognize that, first, when the numbers in a multiplication are positive, a plus sign need not be used; and second, parentheses are necessary only if the numbers are negative.

The rule given in review item 20 relates to what happens when we wish to multiply more than two signed numbers. We could, of course, form the product of two numbers at a time until we had used up all the factors, and then apply the sign we obtained for the last product. However, following the rule simplifies the procedure.

Thus, $(2)(-3)(4)(-5) = +120$ (since there are an even number of negatives), whereas $(-2)(+3)(-4)(-5) = -120$ (since there are an odd number of negatives).

Apply the rule in the following problems.

(a) $(1)(-3)(-2)(-5) =$ ____   (e) $(8)(-8)(-1)$ ____

(b) $(-7)(4)(6)(-2) =$ ____   (f) $(-2)(-2)(-2)$ ____

(c) $(5)(-5)(5) =$ ____   (g) $(+y)(-y)(+y)(-y)$ ____

(d) $(-2)^3 =$ ____   (h) $(-7)(2)(0)(-1) =$ ____

- - - - - - - - - - - - - -

(a) –30; (b) 336; (c) –125: (d) –8; (e) 64; (f) –8; (g) $+y^4$ (h) 0 [remember, from arithmetic, that the product of two or more numbers is zero if any of the numbers (factors) is zero.]

21.   Did you notice anything special about problems (d) and (f) above? They are the same; they are only written differently. However, in its exponential form, $(-2)^3$, the problem gives us a clue to another variation on our multiplication rules. The variation is stated for you in review item 21. Let us see what it means.

Looking at problem (d) in reference item 20, it is apparent that the exponent 3 gives an odd (that is, not even) power of –2, which is a negative number. This means that there will be an *odd* number of negative factors—three, to be exact. According to our rule, therefore, the product should be negative. Thus, $(-2)(-2)(-2) = -8$. If the exponent is an even number, such as 4, then there are an *even* number of negative factors, and the answer is positive. Thus, $(-2)^4 = (-2)(-2)(-2)(-2) = +16$.

Apply the rule for exponents in the following problems:

(a) $(-4)^2 =$ ____   (f) $(-7)^1 =$ ____

(b) $(-3)^2 =$ ____   (g) $(-3)^3 =$ ____

(c) $(-5)^3 =$ ____

(h) $(-7)^2 =$ ____

(d) $(-1)^9 =$ ____

(i) $(-y)^3 =$ ____

(e) $(-a)^4$ ____

(j) $(-5)^4 =$ ____

------- ---------

(a) +16; (b) +9; (c) –125; (d) –1; (e) a^4; (f) –7; (g) –27; (h) +49;
(i) $-y^3$; (j) +625

22. As review item 22 indicates, the rule for determining the sign of the quotient when *dividing* signed numbers is very much like the rule for signs in multiplication. In both multiplication and division, if the two numbers involved have the same sign, the result will be positive; if they have different signs, the result will be negative. For example,

$$\frac{+6}{+3} = +2 \qquad \frac{-6}{-3} = +2 \qquad \frac{+6}{-3} = -2 \qquad \frac{-6}{+3} = -2$$

Use the division rule in solving these problems.

(a) $\dfrac{-4}{-2} =$ ____

(f) $\dfrac{-2}{0} =$ ____

(b) $\dfrac{-9}{+3} =$ ____

(g) $\dfrac{-24}{+8} =$ ____

(c) $\dfrac{+16}{-4} =$ ____

(h) $\dfrac{+120}{-12} =$ ____

(d) $\dfrac{+25}{+5} =$ ____

(i) $\dfrac{+2.5}{-0.5} =$ ____

(e) $\dfrac{-1}{-1} =$ ____

(j) $\dfrac{-15}{-6} =$ ____

------- ---------

(a) +2; (b) –3; (c) –4; (d) +5; (e) +1; (f) undefined (remember that division by zero is meaningless); (g) –3; (h) –10; (i) –5; (j) $+2\frac{1}{2}$

23. Here are a few more practice problems for you.

(a) $\dfrac{(-9)(-2)}{(+6)(+3)} =$ ____

(e) $\dfrac{(+2)(-3)}{(-1)(-4)} =$ ____

(b) $\dfrac{(5)(-5)(+2)}{(-1)(-10)}$ ____

(f) $\dfrac{(-9)(-8)}{(-3)(+4)} =$ ____

(c) $\dfrac{(-4)(+2)(-4)}{(-2)^3} = $ _____   (g) $\dfrac{(4)(5)}{(-2)^2} = $ _____

(d) $\dfrac{(-3)(12)}{(+3)(-3)(2)} = $ _____   (h) $\dfrac{(-6)(+5)(-2)}{(-1)(-12)} = $ _____

- - - - - - - - - - - - - -

(a) 1;   (b) –5;   (c) – 4;   (d) +2;   (e) $-1\frac{1}{2}$;   (f) – 6;   (g) +5;   (h) +5

24. Here are two more examples of the procedure for evaluating expressions containing signed numbers.

	Example No. 1	Example No. 2
Evaluate:	$2x - 5y = ?$	$ab - \dfrac{3b}{a} = ?$
For the values:	$x = -3; y = +2$	$a = 3; b = -4$
Substituting, we get:	$2(-3) - 5(+2) = ?$	$3(-4) - \dfrac{3(-4)}{3} = ?$
Performing multiplications:	$-6 - 10 =$	$-12 - \dfrac{-12}{3} = ?$
Combining terms and dividing:	$= -16$	$-12 + 4 = -8$

Here is a slightly more difficult problem. Evaluate the expression $\dfrac{2ab - 4b^2}{ab}$ for $a = 2$ and $b = -3$.

Substituting, we get,

$$\dfrac{2(2)(-3) - 4(-3)^2}{2(-3)}$$

Applying the exponent:

$$\dfrac{2(2)(-3) - 4(9)}{2(-3)}$$

Performing multiplications

$$\dfrac{-12 - 36}{-6}$$

Combining terms in numerator and dividing $\dfrac{-48}{-6} = 8$

Using the rules and the above examples as a guide, evaluate the following expressions.

(a) $3xy^2$, for $x = 1$, $y = -2$ _____

(b) $2x^2 - y^2$, for $x = -2$, $y = 3$ _____

(c) $ab^2 + 4$, for $a = -3$, $b = -4$ _____

(d) $x^2 + x^3 - x$, for $x = -4$ _____

(e) $\dfrac{ab + 8}{a^2 + b}$, for $a = -3$, $b = -4$ _____

(f) abc^2, for $a = 1$, $b = 2$, $c = -3$ _____

(g) $x^2 - 2xy + y^2$, for $x = -2$, $y = 3$ _____

(h) $b(b^3 - 2)$, for $b = -2$ _____

(i) $\dfrac{2x - 3y}{4z}$, for $x = -2$, $y = -1$, $z = 3 =$.

------ ---------

(a) $3(1)(-2)^2 = 3(4) = 12$;   (b) $2(-2)^2 - (3)^2 = 2(4)-9 = -1$;   (c) $(-3)(-4)^2 + 4$ $= -3(16) + 4 = -48 + 4 = -44$;   (d) $(-4)^2 + (-4)^3 - (-4) = 16 + (-64) + 4 - 20$

$- 64 = -44$;   (e) $\dfrac{(-3)(-4) + 8}{(-3)^2 + (-4)} = \dfrac{(-3)(-4) + 8}{9 - 4} = \dfrac{12 + 8}{5} = 4$;   (f) $+18$;   (g) $+25$;

(h) $+20$;   (i) $-\dfrac{1}{12}$

UNIT THREE

Monomials and Polynomials

Review Item	Ref Page	Example
1. A *polynomial* is a multinomial whose letters (variables) have only positive integers as exponents. It cannot contain a negative or a fractional exponent. Therefore, it cannot contain a variable in the denominator or under a radical sign.	50	$3y^2 - 6$ and $3a^2 - 3a + 7$ are polynomials because they contain no fractional or negative exponents. $2\sqrt{x} + 3y$ and $\dfrac{2}{3x^2} - 7$ are multinomials, but not polynomials.
2. For clarity and convenience, polynomials should be arranged in the order of descending powers of one of the letters (usually the letter that appears most frequently or to the highest power).	51	Rearranging the terms in the order of descending powers of x, $4x - 7 + 2x^2 + 3x^3$ becomes $3x^3 + 2x^2 + 4x - 7$
3. Grouping symbols are commonly used in algebra. (*Note:* These symbols, like most mathematical symbols, have other uses as well.)	52	1. Parentheses: ( ) 2. Brackets: [ ] 3. Braces: { } 4. Bar: ——
4. Parentheses, brackets, braces, and fraction bars have the effect of grouping terms to act as a single multiplier.	52	$(a + b)$, $\dfrac{a + b}{2}$, or, to avoid possible confusion, $\left(\dfrac{a + b}{2}\right)$.

Review Item	Ref Page	Example
5. *Like terms* are those having identical literal parts (the same letters). They may, and often do, have different numerical factors.	52	$2ab$ and $5ab$, $3x^2y$ and $7x^2y$, mk^2 and $4mk^2$ are all pairs of like terms. abc and bcy, x^2y and xy^2 are *not* pairs of like terms.
6. To add like terms, add their numerical coefficients and keep their common literal coefficient.	53	$3ab + 4ab - 2ab = (3 + 4 - 2)ab$ $= 5ab$ $2x^2y + 4x^2y = (2 + 4)\,x^2y$ $= 6x^2y$
7. To simplify an expression containing a mixture of like and unlike terms, add the like terms first; then write down the unlike terms.	53	$2x + a - 5x + 3y + 4 =$ $(2 - 5)x + a + 3y + 4$, or $-3x + a + 3y + 4$
8. To add polynomials, group like terms together, and then combine them. Addition can be performed either horizontally or vertically.	54	$(2x + 4y - 3z) + (3x - 2y + z)$ $= (2x + 3x) + (4y - 2y) +$ $(-3z + z)$ $= 5x + 2y - 2z$; or, adding vertically: $2x + 4y - 3z$ $\underline{3x - 2y + z}$ $5x + 2y - 2z$
9. To subtract a number, letter, or term, add its opposite (see review item 18, Unit 2).	54	$(+3x) - (+4x)$ $= (+3x) + (-4x)$ $= 3x - 4x = -x$
10. To subtract two polynomials, change each sign in the subtrahend to its opposite and proceed as in the addition of polynomials. Subtraction, like addition, can be performed either horizontally or vertically.	56	$(3k + 2m - 4n) - (k - m + n)$ $= (3k + 2m - 4n) + (-k + m - n)$ $= 3k + 2m - 4n - k + m - n$ $= 2k + 3m - 5n$ or, subtracting vertically: $3k + 2m - 4n$ $\underline{k - m + n}$ $2k + 3m - 5n$

Review Item	Ref Page	Example
11. A simpler way of applying the subtraction rule (see review item 10) is to drop the subtraction symbol and change the sign of each term in the subtrahend as you remove the parentheses.	56	$(2x + 3y - 4z) - (x - y + z)$ Removing the parentheses: $2x + 3y - 4z - x + y - z$ $= x + 4y - 5z$
12. When removing grouping symbols in order to simplify an expression, remove one set at a time, beginning with the innermost set and following the rule of signs (see review item 11).	57	$4y - \{7 - [3 + (2y - 5)]\}$ $4y - \{7 - [3 + 2y - 5]\}$ $4y - \{7 - 3 - 2y + 5\}$ $4y - 7 + 3 + 2y - 5$ $6y - 9$
13. To multiply powers of the same base, keep the base and add the exponents.	58	$x^2 \cdot x^5 = x^{2+5} = x^7$ $3^a \cdot 3^b = 3^{a+b}$ $-2k^2 \cdot 3k^3 = -6k^5$
14. To find the power of a power of a base, keep the base and multiply the exponents.	59	$(x^3)^2 = x^6$ $(k^4)^3 = k^{12}$ $(a^2)^3 \cdot a^5 a = a^{12}$
15. To multiply a polynomial by a monomial, multiply each term of the polynomial by the monomial. Apply the distributive law (see review item 15, Unit 1).	59	Multiplying horizontally: $2xy(x + y - 3) = 2x^2y + 2xy^2 - 6xy;$ or, multiplying vertically: $2x + 3y - 4$ <u>$2xy$</u> $4x^2y + 6xy^2 - 8xy$

Review Item	Ref Page	Example
16. To multiply a binomial by a binomial: • Arrange the terms in same order. • Multiply the upper binomial by the lower left and right terms. • Add the like terms.	60	Multiply: $(2x + 3)$ by $(2 + x)$ $2x + 3$ $\underline{x + 2}$ $2x^2 + 3x$ $\underline{ + 4x + 6}$ $2x^2 + 7x + 6$
17. To multiply a trinomial by a binomial, first arrange each expression in descending order of powers. Then multiply each term of one expression by each term of the other.	61	Multiply: $(2x^2 + 3x - 4)$ by $(x + 2)$ $2x^2 + 3x - 4$ $\underline{x + 2}$ Multiply by x: $\;2x^3 + 3x^2 - 4x$ Multiply by 2: $\underline{ + 4x^2 + 6x - 8}$ Add: $\;2x^3 + 7x^2 + 2x - 8$
18. To divide powers: • If the exponent of the numerator is larger than that of the denominator: • If the exponent of the denominator is larger than that of the numerator: • If the exponents of the numerator and denominator are equal (remember, zero cannot be used as a denominator):	61	$\dfrac{x^a}{x^b} = x^{a-b}; \dfrac{x^7}{x^3} = x^{7-3} = x^4$ $\dfrac{x^a}{x^b} = \dfrac{1}{x^{b-a}}; \dfrac{x^3}{x^7} = \dfrac{1}{x^{7-3}} = \dfrac{1}{x^4}$ $\dfrac{x^a}{x^a} = 1; \dfrac{x^3}{x^3} = 1$

Review Item	Ref Page	Example
19. One over a power may also be written using a negative exponent.	62	$\dfrac{1}{k^2} = k^{-2}$; $\dfrac{1}{x^a} = x^{-a}$
20. To divide monomials, follow the rule for dividing powers and numbers.	62	$\dfrac{25a^3}{5a^2} = 5a$; $\dfrac{35x^3y^2}{7xy} = 5x^2y$ (assuming none of the variables is zero)
21. To divide a polynomial by a monomial, divide each term of the polynomial by the monomial.	63	$\dfrac{4x^3 + 8x^2 - 12x}{2x} = 2x^2 - 4x - 6$

UNIT THREE REFERENCES

1. In Unit 1, we reviewed the fact that a monomial is simply an algebraic expression that does not involve addition or subtraction—although it may, and usually does, involve multiplication, division, or both. We also defined a multinomial as the sum of two or more monomials, and binomials and trinomials as multinomials consisting of either two or three monomials, respectively.

Now we are going to talk about a special kind of multinomial, one whose letters or variables have only positive integers as exponents. Such a multinomial is called a *polynomial*. Since the letters in a polynomial can only have "positive integers" as exponents, a polynomial cannot have fractional or negative exponents. It also means that it cannot have a variable (such as x or y) in the denominator, nor can it have a variable under a radical sign (as in $\sqrt{x}$). (You will understand the reasons for these conclusions better when you get to Unit 6.)

The expressions $6x$, $3y^2 - 6$, and $5a^2 - 3a - 7$ are all polynomials. However, the expressions

$$\frac{3}{x}, \frac{x-3}{5x}, \frac{7}{x} - \frac{2}{3x^2}, \text{ and } 2\sqrt{x}$$

are *not* polynomials. Three of these expressions contain a variable (x) in the denominator; one of them contains a variable under the radical sign (the square root of x).

From this point on we will be dealing more with polynomials than with multinomials; therefore, polynomials is the word you will hear more frequently in connection with algebraic expressions of two or more terms.

Indicate which of the following are polynomials and which are simply expressions.

(a) $3n^2k + \dfrac{1}{2}mv^2$ _____

(b) $7xy^2\,(3z)$ _____

(c) $\dfrac{2}{b} - 7$ _____

(d) $y^2 - 3\sqrt{x} + 4$ _____

(e) $k + \dfrac{3}{\sqrt{x}}$ _____

- - - - - - - - - - - - - -

(a) polynomial;   (b) polynomial;   (c) expression (because the variable b occurs in the denominator of the first term);   (d) expression (because the second term contains the variable x under the radical sign);   (e) expression (because the second term contains $\sqrt{x}$)

2. Here is another example illustrating how a polynomial should be rearranged in the order of descending powers of x:

$$x + 4 - 3x^2 + 4x^3 \quad \text{becomes} \quad 4x^3 - 3x^2 + x + 4$$

Rearrange the following polynomials in the correct sequence.

(a) $4b^2 + 3a^2 - 7ab + 14$ (in terms of a) _____

(b) $3xy - y^2 + 2x^3 - 5x^2$ (in terms of x) _____

(c) $3z^2 - 4 - z^2 + 2z$ _____

(d) $abc^2 + ab^2c + a^2bc$ (in terms of a) _____

(e) $7c^2 - 3ac + dc^3 + 11$ (in terms of c) _____

Rearrange these polynomials in order of descending powers, combining terms if possible.

(f) $7 - x^3 + 2x - 4x^2$ _____

(g) $xy - y^2 + 3x^2 - 2$ _____

(h) $4k^2 - 5 - 2k^2 + 7k$ _____

(i) $xy^3 + 2y - 4 - 3y^2$ _____

- - - - - - - - - - - - - -

(a) $3a^2 - 7ab + 4b^2 + 14$;   (b) $2x^3 - 5x^2 + 3xy - y^2$;   (c) $3z^2 - z^2 + 2z - 4$;
(d) $a^2bc + ab^2c + abc^2$;   (e) $dc^3 + 7c^2 - 3ac + 11$;   (f) $-x^3 - 4x^2 + 2x + 7$;
(g) $3x^2 + xy - y^2 - 2$;   (h) $2k^2 + 7k - 5$;   (i) $xy^3 - 3y^2 + 2y - 4$

3.   There are several grouping symbols used in mathematics you should be aware of. Below are examples of the ways in which the various grouping symbols shown in review item 3 are used.

> *Parentheses:* ( ) as in $2(2x - y)$
> *Brackets:* [ ] as in $3 - [2 + (x - y)]$
> *Braces:* { } as in $z - \{4 + [y - (2 + x)]\}$
> *Bar:* ____ as in the fraction $\dfrac{3 - 4x}{y}$

4.   We covered this rule in Unit 1, references items 17 and 18, but we restate it here because of its importance in combining algebraic expressions. For example, in the expression

$$ak^3 - \left(\frac{a - b}{2}\right) + 14k$$

We can drop the parentheses and write $ak^3 - \dfrac{a - b}{2} + 14k$.

5.   Again, this item is included to reinforce a point covered in Unit 1 (see review item 25) because we will be working with it shortly.
  Write down the like terms (if any) in the following polynomials.

(a) $a + 3a - x + z$ _____

(b) $7 - 3y + 3$ _____

(c) $2x^2y + xy^2 - xy$ _____

(d) $3a^2b^2 - 6a^2b^2 + a^2b^2c$ _____

(e) $2ac + 3ab - 4bc + ac$ _____

(f) $2(a + b) + 3(a - b) - (a + b)$ _____

(g) $2xy + yz - 4xyz + 3yz$ _____

(h) $a^2b^2 + a^2 - b^2 + 2a^2$ _____

(i) $xy + xz - yz$ _____

(j) $2(x + y) - 3(x - y) - 4(x + y)$ _____

(a) a and $3a$; (b) 7 and 3; (c) none; (d) $3a^2b^2$ and $-6a^2b^2$; (e) $2ac$ and $+ac$;
(f) $2(a+b)$ and $-(a+b)$; (g) yz and $3yz$; (h) a^2 and $2a^2$; (i) none;
(j) $2(x+y)$ and $-4(x+y)$

6. If the terms you are adding contain parentheses and signs of operation, remember that you may omit the signs of operation and the parentheses and use only the signs of quality (see review item 15, Unit 2). A plus sign before the first term may also be dropped. For example, $(+2y) + (-3y) + (+5y)$ may be written $2y - 3y + 5y$. Adding the numerical coefficients of these like terms and keeping the common literal coefficient gives us $(2 - 3 + 5)y$, or $4y$.

7. Here is another example illustrating how an expression composed of like and unlike terms is simplified: $2a + 3a + 4ay - 4a + 7z - 4 = (2 + 3 - 4)a + 4ay + 7z - 4 = a + 4ay + 7z - 4$.

Try simplifying the following polynomials.

Example: $3xy + 2x^2y - xy + 4x^2y = (3 - 1)xy + (2 + 4)x^2y = 2xy + 6x^2y$

(a) $4ab - 6bc + 3bc - 2ab =$ _____ = _____

(b) $(+3a) - (+2a) + (-4a) =$ _____ = _____

(c) $x^2 - y^2 + 4 - 3y^2 =$ _____ = _____

(d) $rw - 2rw + 7 - 3 + y =$ _____ = _____

(e) $abc + bac + acb =$ _____ = _____

Simplify and add the following monomials.

(f) $2a + 4c - 6a + 2c + b =$ _____

(g) $3xy - y^2 - 2x^2 + 4y - 3x^2 =$ _____

(h) $2x^2 + 3 - 4x^2 - 5x - 2x =$ _____

(i) $2 + 7 - 4 + k =$ _____

------ --------

(a) $(4 - 2)ab + (-6 + 3)bc = 2ab - 3bc$;   (b) $3a - 2a - 4a = -3a$;
(c) $x^2 + (-1 - 3)y^2 + 4 = x^2 - 4y^2 + 4$;   (d) $(1 - 2)rw + (7 - 3) + y = -rw + 4 + y$;
(e) $(1 + 1 + 1)abc = 3abc$ or $3bac$ or $3acb$;   (f) $-4a + 6c + b$;
(g) $-5x^2 + 3xy - y^2 + 4y$;   (h) $-2x^2 - 7x + 3$;   (i) $k + 5$.

8. Here are two more examples of the procedures for adding polynomials.

Example 1: $(3x - y) + (2y + x)$

Removing parentheses and grouping like terms, we get:

$$(3x + x) + (2y - y) = 4x + y$$

Example 2: $(3x^2y + x - 2xy^2 + 4) + (-3x + 2y + 5xy^2 - 7)$

Combining like terms gives us:

$$3x^2y - 2x + 3xy^2 + 2y - 3$$

Or, arranging the terms of the last example in columns containing like terms, we could write

$$
\begin{array}{l}
3x^2y + x \; - \; 2xy^2 + 4 \\
\quad\quad -3x \; + 5xy^2 - 7 + 2y \\
\hline
3x^2y - 2x + 3xy^2 - 3 + 2y
\end{array}
$$

and adding:

gives us:

Use either the horizontal or the vertical procedure to add the following polynomials.

(a) $(3a^2 + ab + c) + (2c - 4a^2 - ab) = $ _____

(b) $(2x - y) + (3y - x) + (3x + 7) = $ _____

(c) $(x^3 + 2x) + (x^2 - 4) + (x - 2x^2) = $ _____

(d) $(10x^2y + 3xy^2 - 3xy) + (5xy + x^2 - y^2) = $ _____

(e) $(4x^2 - 3 + 2x) + (5x - 2x^2) = $ _____

(f) $(x + y) + (2x - y) + (9 + y) = $ _____

(g) $(3a^2b - ab + b^2) + (3b^2 - a^2b + 2ab) = $ _____

(h) $(k^3 + 8) + (2k - 1) + (3k - 4k^3) = $ _____

- - - - - - - - - - - - - -

(a) $-a^2 + 3c$; (b) $4x + 2y + 7$; (c) $x^3 - x^2 + 3x - 4$; (d) $10x^2y + 3xy^2 + 2xy + x^2 - y^2$; (e) $2x^2 + 7x - 3$; (f) $3x + y + 9$; (g) $2a^2b + ab + 4b^2$; (h) $-3k^3 + 5k + 7$

9. Following are two more examples of the rule given in review item 9 as it applies to subtracting monomials.

Example 1: Subtract $-5a$ from $+2a$.

We write this first as $(+2a) - (-5a)$. Then, since $5a$ is the opposite of $-5a$, we write $(+2a) + (+5a)$, from which we get $2a + 5a = 7a$.

Example 2: $(-5x^2) - (+2x^2)$, which we rewrite as $(-5x^2) + (-2x^2)$. Then, simplifying gives us $-5x^2 - 2x^2 = -7x^2$.

Here is an example with more than two terms:

Example 3: $(+7ab) - (3ab) - (-ab) - (-2ab)$

Each minus sign before a parenthesis indicates a subtraction. Therefore, we *add* the opposite of each term in parentheses. For the above problem, this gives us $(+7ab) + (-3ab) + (+ab) + (+2ab)$ or, simply, $7ab - 3ab + ab + 2ab = 7ab$.

Example 4: $(+2bc) - (-3bc) + (-10bc) - (+5bc)$

Changing subtraction to addition, we get:

$(+2bc) + (+3bc) + (-10bc) + (-5bc)$

Finally, we simplify and combine terms:

$2bc + 3bc - 10bc - 5bc = -10bc$

Here are some practice problems for you. Perform the subtractions indicated below.

(a) $(+3k) - (-7k) = $ _____ $=$ _____

(b) $(-2z^2) - (-5z^2) = $ _____ $=$ _____

(c) $(-7x^2y) - (+2x^2y) = $ _____ $=$ _____

(d) $(+11ab) - (+4ab) = $ _____ $=$ _____

Combine the following like terms.

(e) $(-3xy) - (+xy) + (+4xy) - (-5xy) = $ _____

(f) $(+7ak^2) + (-ak^2) - (+3ak^2) - (-2ak^2) = $ _____

(g) $(-2abc) - (-3\ abc) - (-\ 4abc) - (-abc) = $ _____

------ ---------

(a) $(+3k) + (+7k) = 3k + 7k = 10k$;   (b) $(-2z^2) + (+5z^2) = 3z^2$;   (c) $(-7x^2y + (-2x^2y) = -9x^2y$;   (d) $(+11ab) + (-4ab) = 7ab$;   (e) $(-3xy) + (-xy) + (+4xy) + (+5xy) = 5xy$;   (f) $(+7ak^2) + (-ak^2) + (-3ak^2) + (+2ak^2) = 5ak^2$;   (g) $(-2abc) + (+3abc) + (+4abc) + (+abc) = 6abc$

10.   Since subtraction, like addition, can be performed either horizontally or vertically, here is an example of each method as applied to subtraction of polynomials.

Horizontal subtraction: $(5x^2 - 2xy + 3) - (4x^2 + x - 2)$

First we rewrite the problem:

$$(5x^2 - 2xy + 3) + (-4x^2 - x + 2)$$

Then we remove parentheses and combine like terms:

$$5x^2 - 2xy + 3 - 4x^2 - x + 2 = x^2 - 2xy - x + 5$$

Vertical subtraction:

$5x^2 - 2xy + 3$		$5x^2 - 2xy + 3$
Subtract: $4x^2 \qquad - 2 + x$	Add:	$-4x^2 \qquad + 2 - x$
$\overline{x^2 - 2xy + 5 - x}$		$\overline{x^2 - 2xy + 5 - x}$

Use whichever procedure seems easier to you in arranging and subtracting the following polynomials.

(a) $3x - 2y + 6$ from $5x + 4y + 8 =$ _____

(b) $4ab - b^2 + 2a^2$ from $3a^2 + ab - b^2 =$ _____

(c) $x^3 - 2x + y - 3x^2$ from $x^2 - 3x^3 + 4x =$ _____

(d) $mv^2 - 4 + 3av$ from $2av - 3mv^2 - 4 =$ _____

(e) $2a + 3b - 4$ from $6a - b + 2 =$ _____

(f) $3 - z^2 - 4z$ from $9z + 6z^2 - 1 =$ _____

(g) $x - 2y$ from $5y - x^2 + 3x - 9 =$ _____

(h) $2a - 3a^2 + ab$ from $6a^2 - 5a =$ _____

(a) $2x + 6y + 2$;   (b) $a^2 - 3ab$;   (c) $-4x^3 + 4x^2 + 6x - y$;   (d) $-4mv^2 - av$;
(e) $4a - 4b + 6$;   (f) $7z^2 + 13z - 4$;   (g) $-x^2 + 2x + 7y - 9$;
(h) $9a^2 - 7a - ab$

11.   Let us restate this application of the subtraction rule just to make sure it is clear:

- When removing parentheses preceded by a *plus sign,* drop the plus sign and parentheses and *do not change* the signs of the enclosed terms.
- When removing parentheses preceded by a *minus sign,* drop the minus sign and parentheses and *change* the sign of each enclosed term.

Here is an example of each of these situations.

Parentheses preceded by a plus sign:

$$(10a^2 - 3a) + (-5 - 5a^2) = 10a^2 - 3a - 5 - 5a^2 = 5a^2 - 3a - 5$$

Parentheses preceded by a minus sign:

$$(2xy - y^2 + 3) - (2y^2 + xy) = 2xy - y^2 + 3 - 2y^2 - xy = xy - 3y^2 + 3$$

Combination of both situations:

$$(3m - 2k) - (7m + 4k) + (m - k) = 3m - 2k - 7m - 4k + m - k$$
$$= -3m - 7k$$

Remove parentheses and combine like terms in the following.

(a) $(3a - 4b) - (9a + 2b) = $ _____ $= $ _____

(b) $(5xy + 3z) - (z + 2xy) = $ _____ $= $ _____

(c) $(x^2 + y^2) - (2y^2 - x^2) + (w) = $ _____ $= $ _____

(d) $(a - b + c) + (a + b - c) = $ _____ $= $ _____

(e) $(-k^2 + m - 2n) - (n - 2m + 3k^2) = $ _____ $= $ _____

(f) $(3 - a + b) + (7a) - (5 + 3b) = $ _____ $= $ _____

- - - - - - - - - - - - - -

(a) $3a - 4b - 9a - 2b = -6a - 6b$;   (b) $5xy + 3z - z - 2xy = 3xy + 2z$;
(c) $x^2 + y^2 - 2y^2 + x^2 + w = 2x^2 - y^2 + w$;   (d) $a - b + c + a + b - c = 2a$;
(e) $-k^2 + m - 2n - n + 2m - 3k^2 = -4k^2 + 3m - 3n$;
(f) $3 - a + b + 7a - 5 - 3b = 6a - 2b - 2$

12. Here are two more examples of the procedure for removing grouping symbols. Remember that we work from the innermost to the outermost, i.e., inside—out.

Example 1: $2x - \left\{ 3 - [7 + (4x - 5)] \right\}$

Remove parentheses: $2x - \left\{ 3 - [7 + 4x - 5] \right\}$

Remove brackets: $2x - \left\{ 3 - 7 - 4x + 5 \right\}$

Remove braces: $2x - 3 + 7 + 4x - 5$

Combine like terms: $6x - 1$

Example 2: $3z - \left\{ 7w - [8 - (2z - 3 - w)] - 4 \right\}$

Remove parentheses: $3z - \left\{ 7w - [8 - 2z + 3 + w] - 4 \right\}$

Remove brackets: $3z - \left\{ 7w - 8 + 2z - 3 - w - 4 \right\}$

Remove braces: $3z - 7w + 8 - 2z + 3 + w + 4$

Combine like terms: $z - 6w + 15$

Follow the procedures shown in the examples above to simplify the expressions below by removing the grouping symbols.

(a) $4 + \left\{ 2x - [3y + (4 - 5x - 5y)] \right\} =$ _____

(b) $2a - \left\{ 4 - [3b - (5a + 7 - b)] + 2 \right\} =$ _____

(c) $xy - \left\{ 3 - [(2xy + 5) - 5xy] + 4 \right\} =$ _____

(d) $\left\{ [- (2a - 3b) - 6 + 4a] - 2 \right\} + 5b =$ _____

(e) $x^2 y + \left\{ 3 - [4x + (2x^2 y - 7)] \right\} + x =$ _____

- - - - - - - - - - - - - - -

(a) $4 + \left\{ 2x - [3y + 4 - 5x - 5y] \right\} = 4 + \left\{ 2x - 3y - 4 + 5x + 5y \right\} =$
$4 + 2x - 3y - 4 + 5x + 5y = 7x + 2y;$

· (b) $2a - \left\{ 4 - [3b - 5a + 7 + b] + 2 \right\} = 2a - \left\{ 4 - 3b + 5a + 7 - b + 2 \right\} =$
$2a - 4 + 3b - 5a - 7 + b - 2 = -3a + 4b - 13;$

(c) $xy + \left\{ 3 - [2xy + 5 - 5xy] + 4 \right\} = xy - \left\{ 3 - 2xy - 5 + 5xy + 4 \right\} =$
$xy - 3 + 2xy + 5 - 5xy - 4 = -2xy - 2;$

(d) $\left\{ [- 2a + 3b - 6 + 4a] - 2 \right\} + 5b = \left\{ - 2a + 3b - 6 + 4a - 2 \right\} + 5b =$
$-2a + 3b - 6 + 4a - 2 + 5b = 2a + 8b - 8;$

(e) $x^2 y + \left\{ 3 - [4x + 2x^2 y - 7] \right\} + x = x^2 y + \left\{ 3 - 4x - 2x^2 y + 7 \right\} + x =$
$x^2 y + 3 - 4x - 2x^2 y + 7 + x = -x^2 y - 3x + 10$

13.   In review item 24, Unit 1, we defined an exponent as the small numeral placed to the right and slightly above another numeral or a letter to indicate how many times the latter is to be taken as a factor. The rule given in review item 13 tells how to multiply powers of the same base. (A base together with its exponent is called a *power;* thus, x^2 is a power.) When multiplying powers of the same base, we keep the base and add the exponents. Suppose, for example, we wish to multiply the two monomials $(3a)$ and $(-4a^2)$. We proceed as follows:

> Multiply numerical coefficients: $(3)(-4) = -12$
> Multiply literal coefficients: $(a)(a^2) = a^3$
> Multiply the result: $(-12)(a^3) = -12a^3$

Multiply the following monomials, following the procedures shown in the example above.

(a) $c \cdot c^2 =$ _____   (f) $-2a \cdot 7ab \cdot b^2 =$ _____

(b) $b^2 \cdot b \cdot 2b^6 =$ _____   (g) $a^3 \cdot a^6 =$ _____

(c) $(2^a)(2^b) =$ _____   (h) $x^4 \cdot x^6 \cdot x =$ _____

(d) $2a^2(-3ab) =$ _____   (i) $3^a \cdot 3^b \cdot 3^c =$ _____

(e) $x^2x^3x^4 =$ _____   (j) $-2b^2c \cdot 4b^2c(b^3) =$ _____

- - - - - - - - - - - - - -

(a) c^3;   (b) $2b^9$;   (c) 2^{a+b};   (d) $-6a^3b$;   (e) x^9;   (f) $-14a^2b^3$;   (g) a^9; (h) x^{11};   (i) 3^{a+b+c}; (j) $-8b^7c^2$

14. Since x^2 is a power of a base, then $(x^2)^3$ is the power of a power of a base. The wording may be a bit confusing, but the examples shown should make the meaning clear. Thus, $(x^3)^2$ could be written $(x^3)(x^3) = x^{3+3} = x^6$, as discussed in review item 13. But simply multiplying the two exponents in the original expression $(x^3)^2$ (that is, $x^3 \cdot 2$) would produce the same answer, x^6. Hence the rule, as stated in review item 14:

- To find the power of a power of a base, keep the base and multiply the exponents.

Examples: $(4^3)^4 = 4^{12}$; $(a^5)^3 = a^{15}$; $(y^a)^b = y^{ab}$.

Apply the rule to these problems.

(a) $(a^4)^2 =$ _____   (b) $(x^3)^5 =$ _____   (c) $(w^b)^c =$ _____
(d) $(k^4)^4k^3 =$ _____   (e) $(b^5)^2(b^2)^4 =$ _____   (f) $t^3(t^3)^3 =$ _____

- - - - - - - - - - - - - -

(a) a^8;   (b) x^{15};   (c) w^{bc};   (d) k^{19};   (e) b^{18};   (f) t^{12}

15. In reference item 15, Unit 1, we saw from the distributive law for multiplication that $a(b + c) = ab + ac$. Or, using numbers, $6 \cdot 17 = 6(10 + 7) = 6 \cdot 10 + 6 \cdot 7 = 60 + 42 = 102$. In other words, the distributive law tells us that if we wish to multiply the sum of two numbers by a third number, we can multiply each number of the sum by the third number and then add the products. The result will be the same as if we had found the sum of the two numbers first and then multiplied it by the third number.

Why are we concerned with this now? Because in algebra many algebraic expressions are already divided into several parts. This is, in fact, essentially what polynomials are. The distributive law, therefore, provides our rule for multiplication by a polynomial. Below are some further examples.

$$3(a + b - c) = 3a + 3b - 3c$$
$$7(3 + x) = 21 + 7x$$
$$-ak(a + k) = -a^2k - ak^2$$

Follow the rule in performing these multiplications.

(a) $-2(-a + 9) =$ _____

(b) $ry(r - y) =$ _____

(c) $-k^2(ak - 3 + a) =$ _____

(d) $2x^2y(x^2 - xy + 3y) =$ _____

(e) $-d(a - b - c + d) =$ _____

(f) $\frac{a}{2}(2a - 4 + 3b) =$ _____

- - - - - - - - - - - - -

(a) $2a - 18$; (b) $r^2y - ry^2$; (c) $-ak^3 + 3k^2 - ak^2$; (d) $2x^4y - 2x^3y^2 + 6x^2y^2$;

(e) $-ad + bd + cd - d^2$; (f) $a^2 - 2a + \dfrac{3ab}{2}$

16.   Here is another example for you:

Example: Multiply $(4 - 2y)$ by $(y + 3)$.

Arranging terms in same order:

$$-2y + 4$$
$$\underline{y + 3}$$

Multiplying:

$$-2y^2 + 4y$$
$$\underline{\quad - 6y + 12}$$

Adding like terms:

$$-2y^2 - 2y + 12$$

Perform the following multiplications, rearranging terms where necessary.

(a) $(2a - 3b)$ by $(a + b) =$ _____

(b) $(3m + 1)$ by $(4 - m) =$ _____

(c) $(3k - 2m)$ by $(k + 3m) =$ _____

(a) $2a^2 - ab - 3b^2$;   (b) $-3m^2 + 11m + 4$;   (c) $3k^2 + 7km - 6m^2$

- - - - - - - - - - - - -

17. We are just extending a little bit what we covered in review item 16, where we multiplied a binomial by a binomial. Now we are concerned with the procedure for multiplying a trinomial by a binomial.

Example: Multiply $(r^2 - 3r + 5)$ by $(5r - 2)$.

$$r^2 - 3r + 5$$
$$5r - 2$$

Multiplying: $\quad\overline{5r^3 - 15r^2 + 25r}$
$$- 2r^2 + 6r - 10$$
Adding like terms: $\quad\overline{5r^3 - 17r^2 + 31r - 10}$

In performing the following multiplications, remember to first arrange terms in order of descending powers where necessary.

(a) $(2k - 3 + k^2)(k - 4) =$ _____ (d) $(3 + a^2 - 2a)(a + 2) =$ _____

(b) $(x^2 + xy + y^2)(x + y) =$ ___ (e) $(xy - 2x^2 + y^2)(x - y) =$ _____

(c) $(y + k + 2)(y + 1) =$ _____ (f) $(3ab + 2a^2 - 4)(2a + b) =$ _____

- - - - - - - - - - - - - -

(a) $k^3 - 2k^2 - 11k + 12$; (b) $x^3 + 2x^2y + 2xy^2 + y^3$; (c) $y^2 + ky + 3y + k + 2$; (d) $a^3 - a + 6$; (e) $-2x^3 + 3x^2y - y^3$; (f) $4a^3 + 8a^2b + 3ab^2 - 8a - 4b$

18. To make sure this procedure for dividing powers is clear, let us take as an example the division $\dfrac{x^5}{x^3}$. Assuming that $x \neq 0$ (so that the denominator, x^3, is *not* equal to zero), we can write this as

$$\frac{xxxxx}{xxx}$$

With the division written this way, you will recognize at once that the three x's in the denominator will cancel out with (or, more correctly, divide into) three of the x's in the numerator. Thus

$$\frac{\cancel{xxx}xx}{\cancel{xxx}}$$

This leaves only two factors of x, or x^2, in the numerator. Notice that the answer, x^2, would have been the same had we subtracted the exponent of the denominator, 3, from the exponent of the numerator, 5. Thus, $\dfrac{x^5}{x^3} = x^{5-3} = x^2$ ($x \neq 0$).

Similarly, $\dfrac{x^3}{x^5} = x^{3-5} = x^{-2}$ or $\dfrac{1}{x^2}$

Perform the following divisions. Assume there are no zeros in the denominators.

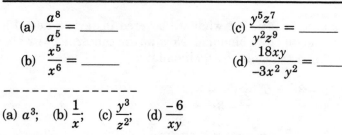

(a) $\dfrac{a^8}{a^5} = $ _____

(b) $\dfrac{x^5}{x^6} = $ _____

(c) $\dfrac{y^5 z^7}{y^2 z^9} = $ _____

(d) $\dfrac{18xy}{-3x^2 \, y^2} = $ _____

- - - - - - - - - - - - -

(a) a^3; (b) $\dfrac{1}{x}$; (c) $\dfrac{y^3}{z^2}$; (d) $\dfrac{-6}{xy}$

19. As we saw in the second example for reference item 18, x^{-2} is the same as $\dfrac{1}{x^2}$. This is true by definition. In general terms, we can say that if $a \neq 0$, then $a^{-n} = \dfrac{1}{a^n}$.

Perform the following divisions of the same base. Assume there are no zeros in the denominators.

(a) $\dfrac{x^8}{x^6} = $ _____

(b) $\dfrac{4^3}{4^5} = $ _____

(c) $\dfrac{a^3}{a^3} = $ _____

(d) $\dfrac{x^{2a}}{x^a} = $ _____

(e) $\dfrac{b^5}{b^7} = $ _____

(f) $\dfrac{-7^3}{7^3} = $ _____

(g) $\dfrac{8^3}{8^7} = $ _____

(h) $\dfrac{m^a}{m^{2a}} = $ _____

- - - - - - - - - - - - -

(a) x^2; (b) $\dfrac{1}{4^2}$ or 4^{-2}; (c) 1; (d) x^a; (e) $\dfrac{1}{b^2}$ or b^{-2}; (f) -1; (g) $\dfrac{1}{8^4}$ or 8^{-4};

(h) $\dfrac{1}{m^a}$ or m^{-a}

20. Here we extend the technique we have been using in the previous two items to the division of monomials in general. Instead of just one exponential factor in the dividend or divisor, there may be several, which may consist of a combination of letters and numbers. For example, consider the division $\dfrac{-48x^3 y^4}{8xy^2}$.

We get our answer by dividing 8 into -48 to get -6, x into x^3 to get x^2, and y^2 into y^4 to get y^2 as the final factor in the quotient $-6x^2 y^2$.

Follow this procedure to solve the following problems. Refer to Unit 2, review item 22, if you are hazy about dividing signed numbers.

(a) $\dfrac{17a^2 y^3}{-17ay} = $ _____

(b) $\dfrac{7abc^2}{14ab} = $ _____

(c) $\dfrac{36xy^3}{6x^2 y} = $ _____

(d) $\dfrac{-28m^3 n}{-7mn^3} = $ _____

---------- ----------

(a) $-ay^2$; (b) $\dfrac{c^2}{2}$; (c) $\dfrac{6y^2}{x}$; (d) $\dfrac{4m^2}{n^2}$

21. The important thing to remember is to divide *each term* of the polynomial by the monomial. Here, again, it will help if you arrange the terms of the polynomial in the descending order of powers of one letter, if they are not already in that order. The fact that the polynomial may be lengthy does not make the problem any more difficult, just longer. Simply take the terms of the polynomial one at a time.

Perform the following division. Assume there are no zeros in the denominators.

Example: $\dfrac{9x^5 - 27x^4 + 18x^3 - 3x^2 + 6x}{3x} = 3x^4 - 9x^3 + 6x^2 - x + 2$

(a) $\dfrac{4k^3 - 2k^2 + 12k - 6}{2} = $ _____

(b) $\dfrac{25a + 15a^3 - 10a^2}{5a} = $ _____

(c) $\dfrac{7m^4n^3 - 14m^5n^4 + 21m^3n^2}{7mn} = $ _____

(d) $\dfrac{9x^2y - 36xy^2}{-9xy} = $ _____

(e) $\dfrac{cd - cdk}{cd} = $ _____

---------- ----------

(a) $2k^3 - k^2 + 6k - 3$;   (b) $3a^2 - 2a + 5$;   (c) $-2m^4n^3 + m^3n^2 + 3m^2n$;
(d) $-x + 4y$;   (e) $1 - k$

UNIT FOUR

Special Products and Factoring

Review Item	Ref Page	Example
1. *Factoring* is expressing a number or algebraic expression as a product of certain factors.	68	Two factors of 21 are 3 and 7, since $3 \cdot 7 = 21$; two factors of $2x^2 - 4x$ are $2x$ and $(x - 2)$, since $2x(x - 2) = 2x^2 - 4x$.
2. Factoring is a special kind of division in which both the divisor and the quotient are to be found.	68	$$\frac{2x^2 - 4x}{2x \text{ (divisor)}} = x - 2 \text{ (quotient)}$$ Both $2x$ and $(x - 2)$ are factors of $2x^2 - 4x$.
3. A *common monomial factor* is the product of the literal and numerical factors common to all terms of a polynomial.	69	$2x^2$ is the common monomial factor of the polynomial $2x^2y^2 - 4$ $x^2y^2 + 6x^2y$. Thus, $2x^2y^2 - 4x^2y^2 + 6x^2y = 2x^2(y^2 - 2y^2 + 3y)$.
4. To square a monomial, square its numerical coefficient, keep each literal factor, and double the exponent of each literal factor.	70	$(4ab^3)^2 = (4ab^3)(4ab^3)$ $$= 16a^2b^6$$
5. The positive square root of a number is called the *principal square root*.	71	The square roots of 9 are $+3$ and -3; $+3$ is the principal square root.

Review Item	Ref Page	Example
6. To find the principal square root of a monomial, find the principal square root of its numerical coefficient, keep each literal factor, and use half the exponent of each literal factor.	71	$\sqrt{36a^4b^6} = 6a^2b^3$
7. To multiply two binomials *by inspection*, or horizontally:	72	$\overbrace{(3x + 2)(x - 1)}$ with $-3x$ above and $2x$ below
• Multiply the first terms.		$(3x)(x) = 3x^2$
• Add the product of the inner terms and the product of the outer terms.		$2x - 3x = -x$
• Multiply the last terms.		$(2)(-1) = -2$
• Combine the terms.		$3x^2 - x - 2$
8. The products of the inner and outer terms are often called the *cross products*.	73	In the example above, $2x$ and $-3x$ are the cross products.
9. To factor a trinomial of the form $x^2 + bx + c$:	74	$x^2 + 5x + 6$ Here $b = 5$, $c = 6$.
• Use x as the first term of each binomial factor.		$(x \qquad)(x \qquad)$
• Use plus signs in the binomials, since these are only plus signs in the trinomials.		$(x + \qquad)(x + \qquad)$
• Use as the second terms of the factors the two numbers whose sum is the second term of the trinomial and whose product is its third term.		The two numbers whose sum is 5 and whose product is 6 are 2 and 3. Thus, $(x + 2)(x + 3)$.

Review Item	Ref Page	Example
	75	
10. To factor a trinomial of the form $ax^2 + bx + c$:	00	$2x^2 + 5x - 18$ Here, $a = 2$, $b = 5$, $c = -18$.
• Since there are no factors of 2 other than 2 and 1, the first terms of the two binomials will be $2x$ and x.		$(2x\quad)(x\quad)$
• Since the last term of the trinomial is minus, the signs of the second terms of the binomials will be opposite.		$(2x +\quad)(x -\quad)$
• The difference of one of the factors of 18, times $2x$, and the other factor of 18, times x, must equal $+5x$, the middle term of the trinomial. 6 and 3 will not work, so we try 2 and 9, which do. (Make sure the larger cross product is positive, as is the second term of the trinomial.)		$(2x + 9)(x - 2)$
11. To factor a polynomial completely, always remove the highest common factor first, and then continue factoring until the polynomial can be factored no further.	76	$4x^3y^2 + 14x^2y^2 + 12xy^2$ $= 2xy^2(2x^2 + 7x + 6)$ $= 2xy^2(2x + 3)(x + 2)$
12. To square a binomial:	77	$(3x - 4)^2$
• Square the first term.		$(3x)^2 = 9x^2$
• Double the product of the two terms.		$2(3x)(-4) = -24x$
• Square the second term.		$(-4)^2 = +16$
• Arrange the terms in descending order.		$9x^2 - 24x + 16$

Review Item	Ref Page	Example
13. Finding the square root of a perfect square trinomial is the reverse of squaring a binomial.	77	Since $(a + b)^2 = a^2 + 2ab + b^2$, $\sqrt{a^2 + 2ab + b^2} = a + b$ (if $a + b$ is greater than zero).
14. The product of the sum and difference of two terms is equal to the square of the first term minus the square of the second term.	78	$(a + b)(a - b) = a^2 - b^2$ $(10 + 3)(10 - 3) = 10^2 - 3^2$ $= 100 - 9$ $= 91$
15. To factor the difference of two squares: • Take the square root of each of the squares. • Write the sum of these square roots as one factor and their difference as the other.	80	Factor: $x^2 - y^2$ $\sqrt{x^2} = x; \sqrt{y^2} = y$ $(x + y)(x - y)$
16. Summary of main points covered in this unit: A polynomial can be factored if: • It contains a monomial factor. • It is a binomial that is the difference of two squares; or • It is a trinomial which is a perfect square or which can be factored by trial and error.	80	 $4a^3 - 6a^2 + 8a = 2a(2a^2 - 3a + 4)$ $a^2 - b^2 = (a + b)(a - b)$ $4m^4 - 9k^2 = (2m^2 + 3k)(2m^2 - 3k)$ $3x^2 + 5x - 12 = (3x - 4)(x + 3)$

UNIT FOUR REFERENCES

1. *Factoring* generally involves finding one factor of a product and then using it as a divisor to find the other factors. It is, therefore, the reverse of multiplication. For example, since $3 \cdot 8 = 24$, we can say that 3 and 8 are factors of the product 24. Or, since $a(a + 4) = a^2 + 4a$, we can say that a and $(a + 4)$ are factors of $a^2 + 4a$.

To make sure there is no confusion in your mind about the relationship among the three processes of multiplication, division, and factoring, study the examples below.

Multiplication	Division	Factoring
factor X factor = product	product ÷ factor = factor	product = factor X factor
$3 \times 7 = 21$	$21 \div 3 = 7$	$21 = 3 \times 7$
$a(a + 3) = a^2 + 3a$	$(a^2 + 3a) \div a = a + 3$	$a^2 + 3a = a(a + 3)$

In the first example given under the head of factoring, it is assumed that we know the product 21 and that we wish to find the two or more factors which, when multiplied together, will produce it. Our procedure, therefore, is to try to divide 21 by various numbers until we finally find one that will divide into it evenly. The number we divide by is one factor, and the quotient is the other factor. If we divide 21 by 3, then 7 is the quotient; if we divide 21 by 7, 3 is the quotient. In either case, 3 and 7 emerge as the factors of 21.

In the second example above, the factoring task consists of finding the component multipliers of the binomial $(a^2 + 3a)$. As you will see (review item 3), the method required involves identifying the common monomial factor—in this case, a. The two factors of $(a^2 + 3a)$ are a and $(a + 3)$. We will discuss this method in more detail later.

2. As you can now see, factoring is a special kind of division in which both the divisor and the quotient are to be found. It is important, then, that you be able to multiply and divide with speed and accuracy if you are to gain any facility in factoring. You should know the following basic principles related to multiplication and division:

• *Two factors of any number are 1 and the number itself.*
Thus, the factors of 9 include 1 and 9; the factors of b include 1 and b, and so on. However, when we speak of factoring a number, we usually mean to find its factor *other than* 1 and the number itself.

• *A prime number is a whole number greater than 1 which has no integral factors except 1 and itself.*

Thus, 2, 3, 7, 11, and 17 are prime numbers. Such numbers as 4, 6, 9, 12, and 15 are *not* prime numbers because they have integral factors other than 1 and themselves; for instance, 9 is divisible by 3, 12 is divisible by 2, 3, 4, and 6, and so on.

• *Monomials need not be factored further, since they already are prime expressions.*

Other polynomials should be factored until they can be factored no further—that is, until all factors are prime expressions.

Now use the following problems to help you practice your multiplication and division skills. (See review item 21, Unit 3, if you need help with the division problems.)

(a) $3(a + b) =$ _____

(b) $2x(x - 3y) =$ _____

(c) $x^2 (1 - 3x) =$ _____

(d) $a^2b(b + 2a) =$ _____

(e) $a(x + y - z) =$ _____

(f) $-5y^3(3x - 2y + 4) =$ _____

(g) $(6x - 3y) \div 3 =$ _____

(h) $(8a^2 + 6a) \div 2a =$ _____

(i) $10k^3 - 4k^2 + 2k) \div 2k =$ _____

(j) $(ab + ac + ad) \div a =$ _____

------ --------- --

(a) $3a + 3b$;   (b) $2x^2 - 6xy$;   (c) $x^2 - 3x^3$;   (d) $a^2b^2 + 2a^3b$;
(e) $ax + ay - az$;   (f) $-15xy^3 + 10y^4 - 20y^3$;   (g) $2x - y$;   (h) $4a + 3$;
(i) $5k^2 - 2k + 1$;   (j) $b + c + d$

3. Factoring a polynomial requires that you first identify the *common monomial factors,* that is, the literal and numerical factors common to all terms of the polynomial. For example, in the expression $3x^2 - 6xy$, 3 and x are the common monomial factors. We then use $3x$ as a divisor and find $(x - 2y)$ to be the quotient. To factor a polynomial, it is important to find both the common factors and the highest common factors.

Example: Find the factors of the polynomial $6x^5 + 12x^4 - 18x^3$.

Solution: Inspection of the three terms reveals that all three numerical coefficients may be divided by 3, and all three literal coefficients by x.

However, 6 is the highest (that is, largest) numerical factor, and x^3 the highest literal factor. Taken together, $6x^3$ represents the *highest common monomial factor* of the polynomial. Dividing by this factor gives us the other factor, $x^2 + 2x - 3$.

Where possible, factor the following.

(a) $2x + 6$ _____

(b) $a + 3$ _____

(c) $3x - 9$ _____

(d) $7 + 3a$ _____

(e) $r + ry - r^2$ _____

(f) $5k^3 - 15k^2 + 35k$ _____

(g) $a^2bc - ab^2c + abc^2$ _____

(h) $6m^3 + 12m^2 - 9m + 1$ _____

(i) $y^3 - 2y + 4$ _____

(j) $10x^3y^2 - 5x^2y^3$ _____

(a) $2(x + 3)$; (b) not possible (prime); (c) $3(x - 3)$; (d) not possible (prime);
(e) $r(1 + y - r)$; (f) $5k(k^2 - 3k + 7)$; (g) $abc(a - b + c)$; (h) not possible
(prime); (i) not possible (prime); (j) $5x^2y^2(2x - y)$

4. Here is another example of the rule given in review item 4:

Example: Square the term $-3x^2y^3$. We write this as $(-3x^2y^3)^2$.

Solution: Squaring the numerical factor, we get $(-3)^2 = 9$. Squaring the literal factors, we get $(x^2y^3)^2 = x^4y^6$. The final result is, therefore, $9x^4y^6$.

Do the following problems. (Remember to square both the numerator and denominator of a fraction.)

(a) $(2mk^2)^2 =$ _____   (d) $(-5a^3b^4)^2 =$ _____

(b) $(-1)^2 =$ _____   (e) $(-ak)^2 =$ _____

(c) $\left(\dfrac{1}{2}gt\right)^2 =$ _____   (f) $\left(\dfrac{2}{5}a^3b\right)^2 =$ _____

- - - - - - - - - - - - - -

(a) $4m^2k^4$;   (b) 1;   (c) $\frac{1}{4}g^2t^2$;   (d) $25a^6b^8$;   (e) a^2k^2;   (f) $\frac{4}{25}a^6b^2$

5. The reverse of squaring a monomial is finding its square root. When a number can be written as the product of two equal factors (such as $25 = 5 \cdot 5$), either of the factors is called the *square root*. Thus, 5 is a square root of 25.

However, since squaring either a positive or negative number produces a positive result, every positive number has two square roots which are equal in absolute value but opposite in sign or quality. Hence the square roots of 25 are $+5$ and -5; more generally, the square roots of a^2 are $+a$ and $-a$. (*Note*: The square root of a negative number is not defined in the real number system.)

The positive square root of a number is called the *principal square root*. The symbol $\sqrt{}$, known as the *radical sign*, is used to indicate the principal (positive) square root of a number. Thus, $\sqrt{64} = 8$. The principal square root of a fraction is the square root of its numerator divided by the square root of its denomina tor. Thus, $\sqrt{\dfrac{25}{49}} = \dfrac{5}{7}$.

The radical sign together with the number under it is called a *radical*. Hence $\sqrt{64}$ and $\sqrt{\dfrac{25}{49}}$ are called radicals. The number under the radical sign is known as the *radicand*. If there is a plus sign or no sign at all before the radical, the positive root is indicated. When there is a minus sign before the radical, the negative root is indicated. Thus, $\sqrt{16} = 4$; $-\sqrt{36} = -6$.

To make sure you have all this straight, supply the missing words in each of the following.

(a) A square root of a number is one of its _____

(b) The positive root is called the _____ square root.

(c) The symbol $\sqrt{}$ is called the _____ sign.

(d) Every positive number has _____ (how many) square roots.

(e) The number under the radical sign is called the _____ .

(f) The combination of the radical sign together with the number under it is known as the _____ .

- - - - - - - - - - - - - -

(a) two equal factors; (b) principal; (c) radical; (d) two; (e) radicand; (f) radical

6. Let us analyze the example from review item 6: Taking the square root of the numerical coefficient, we have $\sqrt{36} = 6$; Keeping each base with half its exponent, $\sqrt{a^4b^6} = a^2b^3$. Combining these, we get $\sqrt{36a^4b^6} = 6a^2b^3$. The problem checks, since $6a^2b^3 \cdot 6a^2b^3 = 36ab^6$.

Further examples:
$$\sqrt{64k^6m^{12}} = 8k^3m^6; \quad \sqrt{81x^2y^2z^2} = 9xyz; \quad \sqrt{\frac{16b^8}{9a^4}} = \frac{4b^4}{3a^2}.$$

Find the principal square roots of the following monomials. Watch your decimal point in problem (d).

(a) $\sqrt{100} =$ _____

(b) $\sqrt{x^2y^6z^4} =$ _____

(c) $\sqrt{\frac{1}{4}k^4m^2} =$ _____

(d) $\sqrt{0.04} =$ _____

(e) $\sqrt{\frac{49}{81}} =$ _____

(f) $\sqrt{\frac{a^4b^6}{c^6d^8}} =$ _____

(g) $\sqrt{\frac{25x^{12}y^{16}}{400z^8}} =$ _____

(h) $\sqrt{81a^2x} =$ _____

(a) 10; (b)xy^3z^2; (c)$\frac{1}{2}k^2m$; (d) 0.2; (e)$\frac{7}{9}$; (f)$\frac{a^2b^3}{c^3d^4}$ (g)$\frac{5x^6y^8}{20z^4}$;
(h) $9a^x$

7. In review item 16, Unit 3, we reviewed the procedure for multiplying two binomials using vertical multiplication. This was an important first step in learning how to multiply larger polynomials. However, in many cases binomials can be multiplied more quickly and simply by means of horizontal multiplication, using the procedure given. This method of multiplying binomials usually is referred to as finding the product *by inspection,* because it is a visual or mental method that enables us to write down the answer directly.

Find the following products by inspection.

(a) $(x + 2)(x - 7) =$ _____

(b) $(2a - 3)(a + 4) =$ _____

(c) $(3k + 4)(3k - 4) =$ _____

(d) $(x + y)(x + y) =$ _____

(e) $(x^2 - 3)(2x^2 + 6) =$ _____

(f) $(4g + 3t)(2g - t) =$ _____

(a) $x^2 - 5x - 14$; (b) $2a^2 + 5a - 12$; (c) $9k^2 - 16$; (d) $x^2 + 2xy + y^2$; (e) $2x^4 - 18$;
(f) $8g^2 + 2gt - 3t^2$

8. Let us perform the multiplication shown in review item 7 by the vertical rather than the horizontal method. You will soon discover where the term *cross products* comes from.

$$
\begin{array}{r}
3x + 2 \\
x - 1 \\
\hline
3x^2 + 2x \\
-3x - 2 \\
\hline
3x^2 - x - 2
\end{array}
$$

The arrows show the "cross multiplication" that occurs in this method; hence the name cross products. They are the result of multiplying the first term of one binomial by the second term of the other, and the second term of one binomial by the first term of the other.

The following exercises will help you develop skill in performing multiplication by inspection. Write in the missing terms.

(a) $(k + 3)(k - 7) = k^2 - 4k - $ _____

(b) $(2a - 3)(a + 2) = $ _____ $+ a - 6$

(c) $(x - 5)(x + 4) = x^2 - $ _____ $- 20$

(d) $(3c + 2)(4c - 5) = $ _____ $- 7c - 10$

(e) $(x^2 - 2)(2x^2 + 1) = 2x^4 - $ _____ -2

(f) $(5 + b)(2 - b) = 10 - $ _____ $- b^2$

Find the products by inspection.

(g) $(2a + 7)(a - 5) = $ _____

(h) $(3 - x)(5 + 2x) = $ _____

(i) $(k^2 - 2)(k^2 + 3) = $ _____

(j) $(x - 2)(x + 2) = $ _____

(k) $(a - 2)(a - 2) = $ _____

(l) $(xy - 3)(xy + 4) = $ _____

— — — — — — — — — — — — — —

(a) 21; (b) $2a^2$; (c) x; (d) $12c^2$; (e) $3x^2$; (f) $3b$; (g) $2a^2 - 3a - 35$;
(h) $15 + x - 2x^2$; (i) $k^4 + k^2 - 6$; (j) $x^2 - 4$; (k) $a^2 - 4a + 4$; (l) $x^2y^2 + xy - 12$

9. We are interested in factoring trinomials because the result of multiplying two binomials usually is a trinomial. Factoring a trinomial is, therefore, just the reverse of multiplying two binomials; it allows us to identify the two original binomials that produced the trinomial. This is often convenient or necessary, as you will see later on. Although the ability to do this kind of factoring easily and quickly involves a certain amount of educated guessing, with a little practice you will find you can become very adept at it.

Here is another example for you.

Example: Factor the trinomial $x^2 - 2x - 24$.

Solution: By examining the first term, we can write at once

$$x^2 - 2x - 24 = (x \qquad)(x \qquad)$$

Now we must find the pair of factors of –24 that has an algebraic sum of –2, the coefficient of the middle term. Since –6 and 4 are the only factors of –24 with a sum of –2, we can write

$$x^2 - 2x - 24 = (x + 4)(x - 6)$$

Factor the following trinomials.

(a) $a^2 + a - 12 =$ _____

(b) $x^2 - 2x - 8 =$ _____

(c) $k^2 - k - 20 =$ _____

(d) $c^2 + 2c - 3 =$ _____

     (*Note*: Here you must factor the third term into 3 and 1 since these are the only factors it has.)

(e) $b^2 + 5b - 14 =$ _____

(f) $x^2 - 5x - 6 =$ _____ (*Tip*: The obvious factors, 3 and 2, will not work.)

(g) $p^4 - p^2 - 30 =$ _____

(h) $q^2 - 5q + 6 =$ _____

------ ---------

(a) $(a + 4)(a - 3)$;   (b) $(x - 4)(x + 2)$;   (c) $k - 5)(k + 4)$;   (d) $(c + 3)(c - 1)$;
(e) $(b + 7)(b - 2)$;   (f $(x - 6)(x + 1)$;   (g) $(p^2 - 6)(p^2 + 5)$;   (h) $(q - 3)(q - 2)$

10. In the trinomials we have factored so far, the numerical coefficient of the first term has always been 1 (understood). It would be convenient if all trinomials were this way. Often, however, they are not so obliging. Therefore, in review item 10, we need to consider how to factor trinomials whose highest power terms have numerical coefficients greater than 1. Perhaps another illustration will help.

Example: Factor $4x^2 - 8x - 21$.

Solution: The numerical coefficient 4 makes this trinomial different from those we previously factored. We cannot immediately write the first term of each binomial, since we do not know which factors of 4 are correct. However, we can start by factoring the literal coefficient of 4, namely x^2, which gives us:

$$4x^2 - 8x - 21 = (x \quad)(x \quad)$$

The minus sign before the last term of the trinomial tells us that the signs of the last terms of the binomial factors are opposite. Therefore, we can write:

$$4x^2 - 8x - 21 = (x + \ )(x - \ )$$

From this point on, it is a matter of trial and error. The challenge is to combine the possible factors of 4 and 21 in such a way as to produce -8 as the algebraic sum of the cross products. A little experimenting gives us:

$$4x^2 - 8x - 21 = (2x + 3)(2x - 7)$$

With the aid of the above example, as well as the one given in the review item, try the following problems. Supply the missing terms and signs as required.

(a) $2x^2 + x - 15 = (2x - \_)(x + \_)$

(b) $3a^2 - 2a - 5 = (3a - \_\_\_)(a + \_\_\_)$

(c) $9k^2 - 6k + 1 = (3k \_\_\_\_)(3k \_\_\_\_)$

(d) $5c^2 + 17c + 14 = (\_\_\_ + 7)(\_\_\_ + 2)$

(e) $6a^2 + a - 5 = (\_\_\_ -5)(\_\_\_ + 1)$

(f) $9y^2 + 3y - 2 = ($ _____ $2)($ _____ $1)$

Factor the following trinomials:

(g) $3x^2 + 8x + 5 =$ _____

(h) $2a^2 - 9a + 4 =$ _____

(i) $7k^2 + 9k + 2 =$ _____

(j) $3x^2 - 2xy - 5y^2 =$ _____

(k) $3m^2 + 11mn - 20n^2 =$ _____

(l) $1 - 2a - 3a^2 =$ _____ (Do not change the order of terms.)

- - - - - - - - - - - - - -

(a) $(2x - 5)(x + 3)$; (b) $(3a - 5)(a + 1)$; (c) $(3k - 1)(3k - 1)$;
(d) $(5c + 7)(c + 2)$; (e) $(6a - 5)(a + 1)$; (f) $(3y + 2)(3y - 1)$;
(g) $(3x + 5)(x + 1)$; (h) $(2a - 1)(a - 4)$; (i) $(7k + 2)(k + 1)$;
(j) $(3x - 5y)(x + y)$; (k) $(3m - 4n)(m + 5n)$; (l) $(1 - 3a)(1 + a)$

11. Here is another example of factoring a polynomial:

$$5a^2b^2 - 15ab^2 + 10b^2 = 5b^2(a^2 - 3a + 2)$$
$$= 5b^2(a - 2)(a - 1)$$

In the following exercises, remove the highest common factor first, and then continue factoring until the polynomial can be factored no further.

(a) $10x^2 + 12x + 2 =$ _____

(b) $14ah^2 + 20ah + 6a =$ _____

(c) $15x^2 - 12x - 3 =$ _____

(d) $6a^3 + 8a^2 + 2a =$ _____

(e) $5x^2y^2 + 10xy^2 + 5y^2 =$ _____

(f) $3ax^2 + 3ax - 18a =$ _____

- - - - - - - - - - - - - -

(a) $2(5x + 1)(x + 1)$; (b) $2a(7h + 3)(h + 1)$; (c) $3(5x + 1)(x - 1)$;
(d) $2a(3a + 1)(a + 1)$; (e) $5y^2(x + 1)(x + 1)$; (f) $3a(x + 3)(x - 2)$

12. In review item 7, we discussed multiplying binomials. However, squaring a binomial deserves some special attention because it occurs frequently and can be done easily once you understand the method. We know, for example, that $(x + y)^2$ means $(x + y)(x + y)$ and that the trinomial product is $x^2 + 2xy + y^2$. Notice that this trinomial product consists of the square of the first term of the binomial, twice the product of both terms, and the square of the second term. Compare this example with the following one:

$$(x - y)^2 = x^2 - 2xy + y^2$$

As you can see, the only difference is the sign of the middle term. Thus, when the sign separating the two terms of the binomial is positive, the middle term of the product is positive; when it is negative, the middle term of the product is negative.

Follow the procedure outlined in review item 12 to square the following binomials.

(a) $(a + 6)^2 =$ _____

(b) $(b - 3)^2 =$ _____

(c) $(2x + 1)^2 =$ _____

(d) $(y + 7)^2 =$ _____

(e) $(n - 5)^2 =$ _____

(f) $(3k + 4)^2 =$ _____

(g) $(3 - 2p)^2 =$ _____

(h) $(p + k)^2 =$ _____

------ ----------

(a) $a^2 + 12a + 36$; (b) $b^2 - 6b + 9$; (c) $4x^2 + 4x + 1$; (d) $y^2 + 14y + 49$; (e) $n^2 - 10n + 25$; (f) $9k^2 + 24k + 16$; (g) $9 - 12p + 4p^2$; (h) $p^2 + 2pk + k^2$

13. A perfect square trinomial is always the square of a binomial. A trinomial is a perfect square if, when arranged either in ascending or descending power of one letter, the first and third terms are positive perfect squares and the middle term is equal to twice the product of the square roots of the first and third terms.

Here are a few more examples of perfect square trinomials:

$$x^2 + 10x + 25 = (x + 5)(x + 5)$$
$$a^2 - 6a + 9 = (a - 3)(a - 3)$$
$$x^2 - 2x + 1 = (x - 1)(x - 1)$$

Notice that each of the above trinomials meets the requirements for a perfect square trinomial. In each case, the first and third terms are perfect positive squares and the middle term is twice the product of the square roots of the first and third terms.

Give the factored form of the trinomials below that are perfect squares. (Be careful; not all are perfect squares.)

(a) $k^2 + 12k + 36 = $ _____

(b) $a^2 - 16a + 64 = $ _____

(c) $x^2 + x + 1 = $ _____

(d) $9x^2 - 6x + 1 = $ _____

(e) $z^2 + 12z - 36 = $ _____

(f) $x^4 - 4x^2 + 4 = $ _____

(g) $b^2 + b + \dfrac{1}{4} = $ _____

(h) $81m^2 + 18m + 1 = $ _____

- - - - - - - - - - - - - -

(a) $(k + 6)^2$;   (b) $(a - 8)^2$;   (c) not a perfect square (middle term is x, not $2x$);   (d) $(3x - 1)^2$;   (e) not a perfect square (the term 36 is negative);

(f) $(x^2 - 2)^2$;   (g) $\left(b + \dfrac{1}{2}\right)^2$;   (h) $(9m + 1)^2$

14.   So far, we have reviewed how to quickly square binomials of the form $(a + b)$ and $(a - b)$ and how to factor perfect square trinomials. But there is one more binomial multiplication you should recall and be adept at performing: $(a + b)(a - b)$. This is often referred to as finding the product of the sum and the difference of two terms. Here is how it works out:

$$
\begin{array}{c}
+ab \\
\downarrow \quad \downarrow \\
(a + b)(a - b) = a^2 - b^2 \\
\uparrow \qquad \qquad \uparrow \\
-ab
\end{array}
$$

Note that the two cross products ($+ab$ and $-ab$) add to zero, thus eliminating the middle term we normally would expect to see in the product. The product of this multiplication is, therefore, a binomial rather than a trinomial. Similarly, $(x + 3)(x - 3) = x^2 - 9$; $(xy - 4)(xy + 4) = x^2y^2 - 16$; or, using numbers instead of letters, $(4 - 2)(4 + 2) = 16 - 4 = 12$.

The last example above shows how subtracting the square of the second term from the square of the first term can be used to simplify some awkward multiplications where the two factors can be expressed as the sum and difference of the same two numbers. For example, 17×23 can be expressed as $(20 - 3)(20 + 3)$. Subtracting the square of the second term from the square of the first term, we get $(400 - 9) = 391$. This is a much faster way of solving the problem than straight multiplication. Although the sum-and-difference concept generally is applied to algebraic expressions rather than numbers, as seen in the example above, it can be useful with numbers.

A few problems of both types are included in the following exercises so that you can practice recognizing multiplications that lend themselves readily to this approach. Multiply the following by inspection.

(a) $(c + 7)(c - 7) =$ _____

(b) $(3y + 4)(3y - 4) =$ _____

(c) $(x + ny)(x - ny) =$ _____

(d) $(9 + 3)(9 - 3) =$ _____ $=$ _____

binomial product

(Subtract mentally the square of the second term from the square of the first term.)

(e) $(2h - k)(2h + k) =$ _____

(f) $(20 + 6)(20 - 6) =$ _____ $=$ _____

binomial product

(g) $(33 \times 27) = ($_____$)($_____$) =$ _____ $=$ _____

  sum   diff.   binomial product

(h) $(1 - z^2)(1 + z^2) =$ _____

(i) $(105)(95) = ($_____$)($_____$) =$ _____ $=$ _____

  sum   diff.   binomial product

(j) $\left(\dfrac{a}{b} + 4\right)\left(\dfrac{a}{b} - 4\right) =$ _____

- - - - - - - - - - - - - - -

(a) $c^2 - 49$;   (b) $9y^2 - 16$;   (c) $x^2 - n^2y^2$;   (d) $81 - 9 = 72$;   (e) $4h^2 - k^2$;

(f) $400 - 36 = 364$;   (g) $(30 + 3)(30 - 3) = 900 - 9 = 891$;   (h) $1 - z^4$;

(i) $(100 + 5)(100 - 5) = 10{,}000 - 25 = 9975$;   (j) $\dfrac{a^2}{b^2} - 16$

We can use your knowledge of how to multiply the sum and difference of two terms to formulate a rule for factoring the difference of two squares. Consider the following procedure:

$$\begin{array}{ccc} \text{multiplying} & \text{product} & \text{factoring} \\ (x + y)(x - y) \;\; = & x^2 - y^2 & (x + y)(x - y) \end{array}$$

This allows us to formulate the procedure for factoring the difference of two squares outlined in review item 15.

Example: Factor $c^2 - 9y^2$.

Solution: Since $\sqrt{c^2} = c$ and $\sqrt{9y^2} = 3y$, then $c^2 - 9y^2 = (c - 3y)(c + 3y)$.

Example: Factor $1 - a^2b^2$.

Solution: Since $\sqrt{1} = 1$ and $\sqrt{a^2b^2} = ab$, then $1 - a^2b^2 = (1 - ab)(1 + ab)$.

Factor the following where possible.

(a) $a^2 - b^2 =$ _____

(b) $16 - y^2 =$ _____

(c) $k^2 + 25 =$ _____

(d) $25x^2 - 64y^2 =$ _____

(e) $36 - a^4 =$ _____

(f) $a^2 - b^4 =$ _____

(g) $1 - 81r^4 =$ _____

(h) $p^4 - m^2n^2 =$ _____

(a) $(a - b)(a + b)$;   (b) $(4 + y)(4 - y)$;   (c) cannot be factored because of the + sign;   (d) $(5x + 8y)(5x - 8y)$;   (e) $(6 - a^2)(6 + a^2)$;   (f) $(a + b^2)(a - b^2)$; (g) $(1 - 9r^2)(1 + 9r^2) = (1 + 3r)(1 - 3r)(1 + 9r^2)$ since $(1 - 9r^2)$ can be further factored into $(1 + 3r)(1 - 3r)$;   (h) $(p^2 + mn)(p^2 - mn)$

16. There are other ways of factoring trinomials, some of which you will find in Unit 9.

UNIT FIVE

Fractions

Review Item	Ref Page	Example
1. Algebraic fractions are basically the same as arithmetic fractions, except that they contain letters and powers of letters as well as numerals.	87	Arithmetic fractions: $$\frac{3}{8}, \frac{5}{6}, \frac{13}{16}$$ Algebraic fractions: $$\frac{2}{x}, \frac{3a^2}{7ab}, \frac{2a-3b}{a^2+b^2}$$
2. When the numerator of a fraction is zero and its denominator is *not* zero, the value of the fraction is zero. A fraction with a zero denominator has no meaning, since division by zero is meaningless.	88	$$\frac{0}{8}=0, \quad \frac{0}{a^2}=0, \quad \frac{0}{ab}=0$$ $$\frac{2a^2b}{0} \text{ is meaningless}$$
3. *Equivalent fractions* are fractions that have the same value, but not the same numerators or denominators. Both terms of a fraction may be multiplied or divided by the same number (except zero) without changing the value of the fraction. This is known as the *fundamental principle of fractions*.	89	$$\frac{2}{3}=\frac{4}{6}=\frac{8}{12}=\frac{16}{24}, \text{ and so on.}$$ See the example above, where numerators and denominators have been repeatedly multiplied by 2— or, from right to left, divided by 2.

Review Item	Ref Page	Example
4. A fraction is in its *simplest form* when the numerator and denominator have no common factor except 1.	90	$\dfrac{1}{2}$ is in its simplest form; $\dfrac{2}{4}$ is not in its simplest form.
To reduce a fraction to its simplest form, divide both numerator and denominator by their greatest common factor (GCF). This is an application of the fundamental principle of fractions.		Reduce: $\dfrac{4x^3y^2}{6xy^3}$ $GCF = 2xy^2$ $\dfrac{4x^3y^2 \div 2xy^2}{6xy^3 \div 2xy^2} = \dfrac{2x^2}{3y}$
5. To reduce a fraction by dividing out common binomial factors, start out by factoring the numerator and the denominator.	90	$\dfrac{x^2 - y^2}{3x^2 - xy - 2y^2}$ $= \dfrac{(x-y)(x+y)}{(3x+2y)(x-y)}$ $= \dfrac{x+y}{3x+2y}$
6. *Do not* attempt to reduce fractions by the following: • Adding the same number to both numerator and denominator • Subtracting the same number from numerator and denominator • Cancelling terms separated by plus or minus signs	91	$\dfrac{2}{3} \neq \dfrac{2+1}{3+1} = \dfrac{3}{4}$ $\dfrac{7}{8} \neq \dfrac{7-2}{8-2} = \dfrac{5}{6}$ $\dfrac{x^2 - \cancel{2x^1} - \cancel{8}^4}{\cancel{2x}_1 - 6_3}$ $\neq x^2 - 4/-2$

Review Item	Ref Page	Example
7. When multiplying fractions, the product of two or more fractions is equal to the product of the numerators divided by the product of the denominators.	91	$\dfrac{3}{4} \cdot \dfrac{5}{6} = \dfrac{15}{24} = \dfrac{5}{8}$ (reduced to lowest terms)
Always start by factoring numerators and denominators wherever possible.		$\dfrac{x^2 + 2x + 1}{3x + 3} \cdot \dfrac{6}{x^2 - 1}$ $= \dfrac{{}^1(x+1)(x+1)^1}{1^3(x+1)_1} \cdot \dfrac{{}^2 6}{(x+1)(x-1)}$ $= \dfrac{2}{x-1}$
Divide the numerator and denominator by any common factors.		$\dfrac{3^1}{8_2} \cdot \dfrac{4^1}{9_3} = \dfrac{1}{6'} \quad \dfrac{2a^2}{3b_1} \cdot \dfrac{6b^2}{7a_1} = \dfrac{4ab^2}{7}$
8. The *reciprocal of a number* is 1 divided by the number.	92	The reciprocal of 7 is $\dfrac{1}{7}$. The reciprocal of x is $\dfrac{1}{x}$. The reciprocal of $-\dfrac{1}{3}$ is -3.
9. The *reciprocal of a fraction* is the fraction inverted.	92	The reciprocal of $\dfrac{7}{8}$ is $\dfrac{8}{7}$. The reciprocal of $\dfrac{1}{4}$ is 4.
10. The product of a number and its reciprocal is 1.	93	$\dfrac{7}{8} \cdot \dfrac{8}{7} = 1$
11. To divide by a fraction, multiply by its reciprocal.	93	$6 \div \dfrac{2}{3} = 6 \cdot \dfrac{3}{2}$ $\dfrac{2}{5} \div \dfrac{3}{4} = \dfrac{2}{5} \cdot \dfrac{4}{3}$

Review Item	Ref Page	Example
12. To divide one algebraic fraction by another, multiply by the reciprocal of the divisor, just as with arithmetic fractions.	93	$\dfrac{a^2 - b^2}{a + b} \div \dfrac{a - b}{2ab}$ $= \dfrac{\cancel{(a+b)}(a-b)}{\cancel{(a+b)_1}} \cdot \dfrac{2ab}{\cancel{(a-b)_1}} = 2ab$
13. To solve an equation for an unknown having a fractional coefficient, multiply both members by the reciprocal of the fractional coefficient.	94	$\dfrac{3}{4}x = 9$ $\dfrac{4}{3} \cdot \dfrac{3}{4}x = \dfrac{4}{3} \cdot 9$ or $x = 12$
14. To add or subtract fractions, change them, if necessary, to equivalent fractions, all of which have the same denominator. Place the sum or difference of the numerators over the lowest common denominator (LCD). Then write the resulting fraction in its simplest form.	94	$\dfrac{a}{3} + \dfrac{3a}{6} - \dfrac{4a}{9}$ $= \dfrac{6a}{18} + \dfrac{9a}{18} - \dfrac{8a}{18} = \dfrac{7a}{18}$ $\dfrac{3x}{x + 2} + \dfrac{x}{x - 2}$ $= \dfrac{3x(x - 2) + x(x + 2)}{x^2 - 4}$ $= \dfrac{4x^2 - 4x}{x^2 - 4}$ $= \dfrac{4x(x - 1)}{(x - 2)(x + 2)}$
15. To find the LCD:	95	Find the LCD for $\dfrac{1}{15}$, $\dfrac{3}{8}$, $\dfrac{5}{12}$, and $\dfrac{5}{18}$.
• Write each denominator as the product of prime factors, using exponents as necessary to represent repeated factors.		$15 = 3 \cdot 5$, $8 = 2^3$, $12 = 2^2 \cdot 3$, $18 = 2 \cdot 3^2$
• Write the product of all the different prime factors. • Use the largest exponent from the first step above for each prime factor. The product is the LCD.		$2 \cdot 3 \cdot 5$ $2^3 \cdot 3^2 \cdot 5 = 360$ (LCD)

Review Item	Ref Page	Example
16. To add two or more fractions having binomial or polynomial denominators: • Factor the denominators. • Find the LCD. • Multiply the two members of each fraction by the quotient of the LDC and the denominator of the fraction. • Combine the numerators and write the result obtained over the LCD. • Reduce the resulting fraction to its lowest terms.	96	$\dfrac{1}{3a-6} + \dfrac{1}{6a} - \dfrac{1}{2a+4}$ $= \dfrac{1}{3(a-2)} + \dfrac{1}{6a} - \dfrac{1}{2(a+2)}$ $6a(a-2)(a+2)$ $= \dfrac{2a(a+2)+(a-2)(a+2)}{6a(a-2)(a+2)}$ $= \dfrac{2a^2+4a+a^2-4-3a^2+6a}{6a(a-2)(a+2)}$ $= \dfrac{10a-4}{6a(a-2)(a+2)}$ $= \dfrac{5a-2}{3a(a-2)(a+2)}$
17. In dividing numbers, if the signs of the numerator and denominator are the same, the quotient will be positive; if the signs are different, the quotient will be negative. (See review item 22, Unit 2.)	96	$\dfrac{8}{4} = 2 \qquad \dfrac{-8}{-4} = 2$ $\dfrac{8}{-4} = -2 \qquad \dfrac{-8}{4} = -2$
18. The sign of a polynomial which appears as the numerator or denominator of a fraction is changed by changing all its signs, or by multiplying it by −1.	97	$\dfrac{a^2-4a}{6+a-a^2} = -\dfrac{a^2-4a}{a^2-a-6}$

Review Item	Ref Page	Example
19. For convenience, numerators and denominators of fractions are usually arranged in descending powers of one letter. The first term is made positive.	98	Reduce $\dfrac{x^2 - 4x}{12 + x - x^2}$ to its lowest terms. Rearranging terms of the denominator, we get: $$\dfrac{x^2 - 4x}{-x^2 + x + 12}$$ Multiplying denominator by -1 we get: $$\dfrac{x^2 - 4x}{x^2 - x - 12}$$ Factoring and dividing by $x - 4$, we get: $$\dfrac{x(x-4)^1}{{}_1(x-4)(x+3)} = \dfrac{x}{x+3}$$
20. A *mixed number* consists of a whole number and a fraction. A mixed number can be changed into a simple fraction by adding the whole number and the fraction.	98	$7\dfrac{3}{8}$ and $a + \dfrac{y}{z}$ are mixed numbers. $$2\dfrac{5}{8} = 2 + \dfrac{5}{8} = \dfrac{16}{8} + \dfrac{5}{8} = \dfrac{21}{8}$$ Or, using an algebraic mixed number $$a - b - \dfrac{a^2}{a+b}$$ $$= \dfrac{a-b}{1} - \dfrac{a^2}{a+b}$$ $$= \dfrac{(a-b)(a+b) - a^2}{a+b}$$ $$= \dfrac{a^2 - b^2 - a^2}{a+b} = -\dfrac{b^2}{a+b}$$
21. To change a fraction into a mixed number, divide the numerator by the denominator.	100	$\dfrac{13}{3} = 4\dfrac{1}{3}$ $\dfrac{a^3 - 2a^2 + a - 7}{a}$ $= a^2 - 2a + 1 - \dfrac{7}{a}$

Review Item	Ref Page	Example
22. To change a fraction with a binomial denominator into a mixed expression, use long division.	100	$\dfrac{x^2 - x - 8}{x + 2}$ $$= x + 2 \overline{\smash{\big)}\begin{array}{r} x \quad - \ 3 + \dfrac{-2}{x-2} \\ x^2 \ - \ x - 8 \\ \end{array}}$$ $$\begin{array}{r} x^2 + 2x \\ \hline -3x - 8 \\ -3x - 6 \\ \hline -2 \end{array}$$
23. To change a complex fraction into a simple fraction: • Multiply both numerator and denominator by the LCD of all fractions appearing in them, and reduce as necessary. • Combine terms in the numerator and in the denominator, and then divide the numerator by the denominator.	102	$\dfrac{\dfrac{3}{4}}{\dfrac{1}{3}} = \dfrac{\dfrac{3}{4} \cdot 12}{\dfrac{1}{3} \cdot 12} = \dfrac{9}{4}$ or $2\dfrac{1}{4}$ $\dfrac{a + \dfrac{a}{b}}{\dfrac{1}{b} + \dfrac{1}{b^2}} = \dfrac{ab + a}{b} \div \dfrac{b + 1}{b^2}$ $= \dfrac{a(b+1)^1}{b_1} \cdot \dfrac{b^2 b}{(b+1)_1} = ab$

UNIT FIVE REFERENCES

1. From arithmetic, you know that a fraction is part of a whole. Fractions enable us to represent numerically the fact that we do not have exactly one or two or more complete objects. We need fractions in algebra for the same reason: to represent a part of a whole. More importantly, however, we need fractions to guarantee that the answer to any division will be a usable number. We could not do this if we were limited to the set of integers which, as you will recall from Unit 2, includes only the positive and negative whole numbers and zero.

There is, for example, no way of representing the division $\dfrac{3}{4}$ by any single integer.

Thus, fractions are necessary. The set of integers together with the set of fractions constitutes the set of *rational numbers*. This set includes all the numbers that can be formed by dividing any integer by any other integer other than zero.

The terms of a fraction are the *numerator* and the *denominator*. Thus, in the fraction $\frac{a}{b}$, which we read as "*a* divided by *b*," *a* is the numerator and *b* the denominator. The short horizontal line that separates the numerator from the denominator is called the *fraction bar* (or, sometimes, the *vinculum*). And, as we discussed in review item 2 of Unit 3, in addition to indicating a division, the fraction bar may also serve as a sign of grouping. Remember, too, from review item 9 of Unit 1, that the colon (:) and the arithmetic division symbol (÷) also can be used to indicate division. A ratio (such as 3 to 5) also may be written in fractional form as $\frac{3}{5}$.

To make sure you are clear about the main points covered above, complete the following statements.

(a) The three parts of a fraction are the ——————— , the ——————— , and the ——————— .

(b) Write the three symbols used to indicate division:——, ——, and ——.

(c) A fraction may mean: ——————— , ——————— , or ——— ——— .

- - - - - - - - - - - - - -

(a) numerator, denominator, fraction bar; (b) ——, ÷, and :; (c) division, ratio, part of a whole

2.   As discussed previously (review item 11 Unit 1), division of any number by zero is meaningless. Zero can appear in the numerator of a fraction, in the denominator, or both. The rules governing each of these cases are as follows:

Rule 1:   When the numerator of a fraction is zero, the value of the fraction is zero (provided the denominator is not zero). Thus,

$$\frac{0}{4} = 0, \frac{x}{7} = 0 \qquad (\text{if } x = 0)$$

$$\frac{x-3}{5} = 0 \qquad (\text{if } x = 3)$$

Rule 2: $\frac{0}{0}$ is indeterminate, because the quotient does not exist as a unique number.

Rule 3:   Since division by zero is undefined, a fraction with a zero denominator is meaningless. These fractions are meaningless:

$$\frac{5}{0} \cdot \frac{7}{y}, \frac{a+2b}{y-3} \qquad (\text{where } y = 3)$$

Use these rules to evaluate the following fractions.

(a) $\dfrac{0}{7} =$ _____

(b) $\dfrac{y-7}{4}$ (when $y = 7$) = _____

(c) $\dfrac{13}{z-6}$ (when $z = 6$) = _____

(d) $\dfrac{x-4}{x-8}$ (when $x = 8$) = _____

(e) For what value of x does $\dfrac{x-4}{2+x} = 0$? $x =$ _____

(f) For what value of x is $\dfrac{x-4}{2+x}$ meaningless? $x =$ _____

(g) For what value of y is $\dfrac{y^2+7}{y-5}$ meaningless? $y =$ _____

- - - - - - - - - - - - - -

(a) zero; (b) zero; (c) meaningless; (d) meaningless; (e) 4; (f) –2; (g) 5

3. Two important ideas are involved in this matter of equivalent fractions. They are expressed in the following two definitions:

Definition 1: $\dfrac{a}{c} = \dfrac{b}{d}$ if and only if $ad = bc$. For example, to determine if the two fractions $\dfrac{3}{4}$ and $\dfrac{6}{8}$ are equivalent, cross multiply (3 times 8 and 4 times 6). The result, $24 = 24$, would prove their equivalence.

Definition 2: $\dfrac{a}{b} = \dfrac{ak}{bk} = \dfrac{a/k}{b/k}$, provided b and k are not zero. (By convention, k stands for any number used as a common multiplier or divider for both terms of a fraction.) Thus, referring to the example under Definition 1 above, dividing the larger numerator (6) by the smaller numerator (3) and the larger denominator (8) by the smaller denominator (4) would show they have the same common multiplier, 2. This would prove their equivalence according to Definition 2.

This second definition contains an important idea called the *fundamental principle of fractions*. Notice that the value of a fraction is unchanged if both terms of that fraction are multiplied or divided by the same nonzero number.

Use this principle to determine how the second fraction was obtained from the first in the problems below.

(a) $\dfrac{8}{12} = \dfrac{2}{3}$ (note that $8 \cdot 3 = 12 \cdot 2$) _____

(b) $\dfrac{a}{b} = \dfrac{3a}{3b}$ _____

(c) $\dfrac{x^2}{xy} = \dfrac{x}{y}$ (note that $x^2 \cdot y = xy \cdot x$ _____

(d) $\dfrac{3b - 3d}{6} = \dfrac{b - d}{2}$ _____

- - - - - - - - - - - - -

(a) both terms were divided by 4;   (b) both terms were multiplied by 3;
(c) both terms were divided by x;   (d) both terms were divided by 3

4.   Here is another example of the procedure for reducing a fraction to its
simplest form by dividing both numerator and denominator by their greatest
common factor (GCF). In the fraction

$$\dfrac{8a^2}{12a^3z}$$

4 is the largest numerical factor, and a^2 the largest literal factor. Therefore,
the GCF of both the numerator and denominator is $4a^2$. Dividing both terms
by this factor, we get

$$\dfrac{(8a^2y) \div (4a^2)}{(12a^3z) \div (4a^2)} = \dfrac{2y}{3az}$$

The simplest form of a fraction is also called its *lowest terms*.
   Reduce the following fractions to lowest terms.

(a) $\dfrac{3xy}{9xz} = $ _____

(b) $\dfrac{6ac}{15c^3} = $ _____

(c) $\dfrac{36c^4d^6}{48c^3d} = $ _____

- - - - - - - - - - - - -

(a) $\dfrac{y}{3z}$; (b) $\dfrac{2a}{5c^2}$; (c) $\dfrac{3cd^5}{4}$

5.   Another example of the procedure for reducing a fraction having common
binomial factors is as follows:

Example: Reduce $\dfrac{3x + 3y}{x^2 - y^2}$ to its lowest terms.

Solution: We first factor both numerator and denominator into their prime
           factors:

$$\frac{3(x + y)}{(x + y)(x - y)}$$

We then divide both numerator and denominator by their GCF, $(x + y)$:

$$\frac{3(x + y) \div (x + y)}{(x + y)(x - y) \div (x + y)} = \frac{3}{x - y}$$

Follow this procedure in reducing the following fractions to their lowest terms.

(a) $\dfrac{32c^3d^3}{64c^2d} = $ _____

(b) $\dfrac{5a - 25}{15a} = $ _____

(c) $\dfrac{27x^2}{18x^2 - 9xy} = $ _____

(d) $\dfrac{2x + 6}{3ax + 9a} = $ _____

(e) $\dfrac{a^2 + a}{2 + 2a} = $ _____

(f) $\dfrac{y^2 + 2y - 15}{2y^2 - 12y + 18} = $ _____

- - - - - - - - - - - - - -

(a) $\dfrac{cd^2}{2}$;   (b) $\dfrac{a - 5}{3a}$;   (c) $\dfrac{9 \cdot 3 \cdot x \cdot x}{9 \cdot x(2x - y)} = \dfrac{3x}{2x - y}$; (d) $\dfrac{2(x + 3)}{3a(x + 3)} = \dfrac{2}{3a}$;

(e) $\dfrac{a}{2}$; (f) $\dfrac{y + 5}{2(y - 3)}$

6. Since we do not want to reinforce errors, we will not dwell any further on *mistaken* procedures. Just be aware of the common errors covered in review item 6, so that you will not make any of them inadvertently.

7. Using the examples in the review item as a guide, perform the following multiplications. In problems (g) and (h), be sure to factor the trinomials before attempting to divide out the common factors.

(a) $\dfrac{3}{5} \cdot \dfrac{5}{9} = $ _____

(b) $\dfrac{2x}{y} \cdot \dfrac{x}{2y} = $ _____

(c) $\dfrac{4a}{7} \cdot \dfrac{1}{2b^2} = $ _____

(d) $\dfrac{x^2 y^2}{m^3 n} \cdot \dfrac{m}{x^3} = $ _____

(e) $\dfrac{d}{a+b} \cdot \dfrac{5a+5b}{2d^2+2d} = $ _____

(f) $\dfrac{a+3}{a-5} \cdot \dfrac{2a-10}{3a+9} = $ _____

(g) $\dfrac{a^2-6a+5}{a+1} \cdot \dfrac{a+1}{a-5} = $ _____

(h) $\dfrac{a^2+6a+9}{8} \cdot \dfrac{4a+8}{a^2+5a+6} = $ _____

- - - - - - - - - - - - - -

(a) $\dfrac{1}{3}$;   (b) $\dfrac{x^2}{y^2}$;   (c) $\dfrac{2a}{7b^2}$;   (d) $\dfrac{y^2}{m^2nx}$;   (e) $\dfrac{5}{2(d+1)}$;   (f) $\dfrac{2}{3}$;

(g) $a-1$;   (h) $\dfrac{a+3}{2}$

8.   The concept of the reciprocal of a number is based on the axiom that if a is a real non-zero number, there exists a real number $\dfrac{1}{a}$, called the *reciprocal* of a, such that $a \cdot \dfrac{1}{a} = \dfrac{1}{a} \cdot a = 1$.

9.   The fact that the reciprocal of a number is 1 divided by the number means that the reciprocal of a fraction is the fraction inverted (i.e., turned upside down).

Example: Find the reciprocal of $\dfrac{3}{4}$

Solution: Since the reciprocal must be 1 divided by the number, we write $1 \div \dfrac{3}{4}$. To divide fractions, we invert and multiply (see review item 11 below). This gives us $1 \cdot \dfrac{4}{3} = \dfrac{4}{3}$.

Give the reciprocals of each of the following.

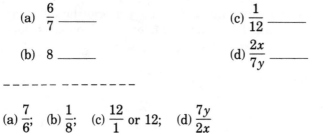

(a) $\dfrac{6}{7}$ _____   (c) $\dfrac{1}{12}$ _____

(b) 8 _____   (d) $\dfrac{2x}{7y}$ _____

- - - - - - - - - - - - - -

(a) $\dfrac{7}{6}$;   (b) $\dfrac{1}{8}$;   (c) $\dfrac{12}{1}$ or 12;   (d) $\dfrac{7y}{2x}$

10. Supply the missing term or product.

(a) $\dfrac{3}{5} \cdot \dfrac{5}{3} =$ ____

(b) $\dfrac{x}{5} \cdot \dfrac{5}{x} \cdot \dfrac{3}{4} =$ ____

(c) $\dfrac{a}{7}(\ ) = 1$

(d) $(\ )\dfrac{1}{3} = 1$

- - - - - - - - - - - - - - -

(a) 1;   (b) $\dfrac{3}{4}$;   (c) $\dfrac{7}{a}$;   (d) 3

11. Change each division to a multiplication by using the reciprocal. Then find the product.

(a) $12 \div \dfrac{3}{4} =$ _____

(b) $\dfrac{a}{2} \div y =$ _____

(c) $\dfrac{a}{b} \div \dfrac{2a}{b} =$ _____

(d) $\dfrac{x}{3} \div \dfrac{3}{x} =$ _____

- - - - - - - - - - - - - -

(a) $12\left(\dfrac{4}{3}\right) = 16$;   (b) $\dfrac{a}{2} \cdot \dfrac{1}{y} = \dfrac{a}{2y}$;   (c) $\dfrac{a}{b} \cdot \dfrac{b}{2a} = \dfrac{1}{2}$;   (d) $\dfrac{x}{3} \cdot \dfrac{x}{3} = \dfrac{x^2}{9}$

12. This is essentially the same rule presented in review item 11, but now we are concerned with somewhat more complex algebraic fractions, such as the ones below.

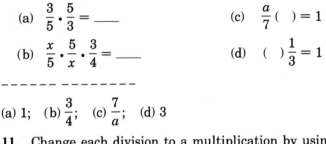

Example: $\dfrac{4a^2}{3ab} \div \dfrac{2a}{6b^2} = \dfrac{\overset{2}{\cancel{4a^2}}}{\cancel{3ab}} \cdot \dfrac{\overset{2}{\cancel{6b^2}}}{\cancel{2a}} = 4b$

Example: $\dfrac{4x^2 - 1}{2x - 6} \div \dfrac{2x^2 - 7x - 4}{x^2 - 7x + 12} = \dfrac{4x^2 - 1}{2x - 6} \cdot \dfrac{x^2 - 7x + 12}{(2x^2 - 7x - 4)} =$

$\dfrac{(2x - 1)\overset{1}{\cancel{(2x + 1)}}}{2\cancel{(x - 3)}} \cdot \dfrac{\overset{1}{\cancel{(x - 4)}}\overset{1}{\cancel{(x - 3)}}}{\underset{1}{\cancel{(2x + 1)}}\underset{1}{\cancel{(x - 4)}}} = \dfrac{2x - 1}{2}$

Now it is your turn. Perform the indicated divisions.

(a) $\dfrac{7}{8} \div \dfrac{3}{4} =$ _____

(b) $\dfrac{a}{n} \div b =$ _____

(Remember: b is the same as $\dfrac{b}{1}$; hence the reciprocal of b is $\dfrac{1}{b}$.)

(c) $\dfrac{2a}{3b} \div \dfrac{6a^2}{b^2} =$ _____

(d) $\dfrac{8}{x^3} \div \dfrac{12}{x^2} =$ _____

(e) $\dfrac{ab}{(a-b)^2} \div \dfrac{1}{a-b} = $ _____

(f) $\dfrac{a^2-9}{a^2+3a} \div \dfrac{a-3}{4} = $ _____

(g) $\dfrac{x^2+7x+6}{x^2+6x+5} \div \dfrac{x^3+6x^2}{x^2+5x} = $ _____

(h) $\dfrac{x^4-y^4}{x+y} \div \dfrac{x^2-y^2}{4x+4y} = $ _____

- - - - - - - - - - - - - -

(a) $\dfrac{7}{6}$; (b) $\dfrac{a}{nb}$; (c) $\dfrac{b}{9a}$; (d) $\dfrac{2}{3x}$; (e) $\dfrac{ab}{a-b}$; (f) $\dfrac{4}{a}$; (g) $\dfrac{1}{x}$; (h) $4(x^2+y^2)$

13. Use the reciprocals to solve these equations.

(a) $\dfrac{2}{3}y = 4$, $y = $ _____   (c) $1\dfrac{2}{3}k = 5$, $k = $ _____

(b) $\dfrac{1}{3}a = 2$, $a = $ _____   (d) $\dfrac{7}{4}n = 14$, $n = $ _____

- - - - - - - - - - - - - -

(a) $\dfrac{3}{2} \cdot \dfrac{2}{3}y = \dfrac{3}{2} \cdot 4$, $y = 6$; (b) $\dfrac{3}{1} \cdot \dfrac{1}{3}a = \dfrac{3}{1} \cdot 2$, $a = 6$; (c) $\dfrac{3}{5} \cdot \dfrac{5}{3}k = \dfrac{3}{5} \cdot 5$, $k = 3$;

(d) $\dfrac{4}{7} \cdot \dfrac{7}{4}n = \dfrac{4}{7} \cdot 14$, $n = 8$

14. In symbols, the adding and subtracting (combining) rule for fractions can be stated as:

$$\dfrac{a}{b} + \dfrac{c}{b} = \dfrac{a+c}{b}$$

Thus, $\dfrac{2}{8} + \dfrac{3}{8} = \dfrac{5}{8}$, and $\dfrac{x}{4} + \dfrac{2x}{8} = \dfrac{2x}{8} + \dfrac{2x}{8} = \dfrac{4x}{8} = \dfrac{x}{2}$. Study these additional examples; then work the problems that follow.

Example: Simplify $\dfrac{x}{2y} - \dfrac{4}{y^2}$; LCD = $2y^2$.

Solution: Multiplying the numerator and denominator of the first fraction by y and the numerator and denominator of the second fraction by 2 gives a common denominator in *both* fractions. The equivalent fractions can then be combined.

$$\dfrac{xy}{2y^2} - \dfrac{2 \cdot 4}{2y^2} \text{ or } \dfrac{xy-8}{2y^2}$$

Example: Combine $\dfrac{2a+1}{2} + \dfrac{a-3}{4} - \dfrac{2a}{3}$; LCD $= 1 \cdot 2 \cdot 3 = 12$.

Solution: $\dfrac{6(2a+1)}{12} + \dfrac{3(a-3)}{12} - \dfrac{4(2a)}{12} = \dfrac{6(2a+1) + 3(a-3) - 4(2a)}{12} =$

$$\dfrac{12a + 6 + 3a - 9 - 8a}{12} = \dfrac{7a - 3}{12}$$

Practice these procedures by combining terms in the following problems.

(a) $\dfrac{2b}{3} - \dfrac{b}{2} + \dfrac{b}{4} =$ _____

(b) $\dfrac{5a}{6} - \dfrac{5a}{12} - \dfrac{a}{3} =$ _____

(c) $\dfrac{4}{n^2} + \dfrac{3}{n} + \dfrac{5}{n^3} =$ _____

(d) $\dfrac{3k+5}{5} - \dfrac{k+3}{3} =$ _____

(e) $\dfrac{2a-5}{14a^2} - \dfrac{4-a}{7a} + \dfrac{3a-2}{2a} =$ _____

- - - - - - - - - - - -

(a) $\dfrac{5b}{12}$; (b) $\dfrac{a}{12}$; (c) $\dfrac{4n + 3n^2 + 5}{n^3}$

(d) $\dfrac{3(3k+5) - 5(k+3)}{15} = \dfrac{9k + 5 - 15k - 15}{15} = \dfrac{4k}{15}$;

(e) $\dfrac{(2a-5) - 2a(4-a) + 7a(3a-2)}{14a^2} = \dfrac{2a - 5 - 8a + 2a^2 + 21a^2 - 14a}{14a^2} =$

$= \dfrac{23a^2 - 20a - 5}{14a^2}$

15. In most of the problems we have encountered so far it was possible to find the LCD either by inspection or by combining denominators. Because this is not always easy to do, we have given you the general procedure set forth in review item 15. Let us work our way through another example.

Consider the fractions $\dfrac{1}{12}, \dfrac{2}{15}, \dfrac{5}{9}$, and $\dfrac{4}{21}$. What is their lowest common denominator? We certainly do not want to multiply 12 · 15 · 9 · 21. Such a product would not be the lowest common denominator because 12, 15, 9, and 21 are not prime numbers. So, following the first step of the rule, we factor all the denominators into their prime factors:

1. (a) 12 written in terms of its prime factors would be _____

   (b) 15 written in terms of its prime factors would be _____

(c) 9 written in terms of its prime factors would be _____

(d) 21 written in terms of its prime factors would be _____

2. Notice that the numbers 12, 15, 9, and 21 actually are composed only of the factors 2, 3, 5, and 7. Therefore, writing the product of all the different prime factors, we get _____

3. Finally, using the largest exponent required for any given prime factor gives us _____ = _____ (LCD).

- - - - - - - - - - - - - -

1. (a) $12 = 2^2 \cdot 3$;   (b) $15 = 5 \cdot 3$;   (c) $9 = 3^2$;   (d) $21 = 3 \cdot 7$;   2. $2 \cdot 3 \cdot 5 \cdot 7$;
3. $2^2 \cdot 3^2 \, 5 \cdot 7 = 1260$ (LCD)

16. In addition to monomial denominators, there may be binomial or other polynomial denominators which you will have to factor before you can determine the LCD.

Follow the step-by-step procedure shown in your review item in working the following problems.

(a) $\dfrac{d}{a + d} - \dfrac{a}{a - d} =$ _____

(b) $\dfrac{3a}{a + 3} + \dfrac{a^2 + 4a - 5}{a^2 - a - 12} =$ _____

(c) $\dfrac{a^2 + b^2}{a^2 - b^2} + \dfrac{2a}{a + b} =$ _____

- - - - - - - - - - - - - -

(a) $\dfrac{d(a - d) - a(a + d)}{(a + d)(a - d)} = \dfrac{ad - d^2 - a^2 - ad}{(a + d)(a - d)} = \dfrac{-a^2 + d^2}{a^2 - d^2}$;

(b) $\dfrac{3a(a - 4) + a^2 + 4a - 5}{(a - 4)(a + 3)} = \dfrac{3a^2 - 12a + a^2 + 4a - 5}{(a - 4)(a + 3)} =$

$\dfrac{4a^2 - 8a - 5}{(a - 4)(a + 3)}$;

(c) $\dfrac{a^2 + b^2 + 2a(a - b)}{(a - b)(a + b)} = \dfrac{a^2 + b^2 + 2a^2 - 2ab}{(a - b)(a + b)} = \dfrac{3a^2 - 2ab + b^2}{(a - b)(a + b)}$

17. This rule works very well where we are concerned with two signs only, that is, the sign of the numerator and the sign of the denominator. In dealing with fractions, however, we have to remain aware of three signs: that of the numerator, that of the denominator, and that of the fraction itself. When any of these signs is omitted, it is understood to be plus. For example:

$$\frac{3}{4} \text{ means } +\frac{+3}{+4}$$

Now consider the four fractions below.

$$+\frac{+6}{+2} = +(+3) = 3 \qquad\qquad -\frac{-6}{+2} = -(-3) = 3$$

$$+\frac{-6}{-2} = +(+3) = 3 \qquad\qquad -\frac{+6}{-2} = -(-3) = 3$$

Notice that the value of each fraction is 3 and that each of the four fractions may be obtained from any of the others by changing two of the three signs. This tells us:

Any two of the three signs of a fraction may be changed without changing the value of the fraction.

State whether each of the following is true or false.

(a) $+\dfrac{-6}{2} = -\dfrac{+6}{+2}$ _____   (c) $-\dfrac{-5}{-7} = \dfrac{-5}{-7}$ _____

(b) $\dfrac{-x}{-y} = \dfrac{x}{y}$ _____   (d) $\dfrac{a}{b} = -\dfrac{-a}{b}$ _____

- - - - - - - - - - - - -

(a) true—the answer is still negative since two signs were changed;   (b) true—exactly two signs were changed;   (c) false—the answer is now positive because only the sign of the fraction was changed;   (d) true—exactly two signs were changed

18.   The fractions in reference item 17 contained monomial numerators and denominators. Now we consider how the rule of the signs of a fraction applies where binomials, trinomials, or polynomials of more than three terms are involved.

Suppose we wish to change the fraction $\dfrac{a - b}{c - d}$ to an equal fraction having the denominator $d - c$. This means that the denominator $c - d$ will have the signs of both its terms reversed. To do this, we multiply the denominator by –1, which gives us the required $d - c$. But to make the new fraction equal the original fraction, we must either change the sign of the fraction or multiply the numerator by –1. Let us choose to multiply the numerator by –1. This gives us:

$$\frac{a - b}{c - d} = +\frac{b - a}{d - c} \quad \text{or} \quad -\frac{a - b}{d - c} \text{ (changing the sign of the fraction)}$$

Remember: The sign of a polynomial is changed by changing the sign of all its terms, or by multiplying it by –1.

Supply the missing terms in the changed fractions below.

(a) $\dfrac{a-3}{b-2} = + \dfrac{?}{2-b}$

(d) $\dfrac{4}{a-b} = + \dfrac{?}{b-a}$

(b) $\dfrac{c-d}{x-y} = - \dfrac{c-d}{?}$

(e) $\dfrac{x-y}{b} = \dfrac{?}{-b}$

(c) $\dfrac{4-a}{x-y} = + \dfrac{?}{y-x}$

(f) $\dfrac{6}{2-y} = \dfrac{?}{y-2}$

- - - - - - - - - - - - -

(a) $3-a$;   (b) $y-x$;   (c) $a-4$;   (d) -4;   (e) $y-x$;   (f) -6

19. Follow the general procedure shown in the example to simplify the fractions below.

(a) $\dfrac{4m-4n}{n^2-m^2} =$ _____

(b) $\dfrac{x^2-4x+3}{3-x} =$ _____

(c) $\dfrac{1-k^2}{3k^2-6k+3} =$ _____

(d) $\dfrac{x^2+x-20}{35+12x+x^2} =$ _____

(e) $\dfrac{(a-b)^2}{b^2+2ba-3a^2} =$ _____

(f) $\dfrac{x^2-6x+9}{9-x^2} =$ _____

- - - - - - - - - - - - -

(a) $-\dfrac{4}{m+n}$;   (b) $-(x-1) = 1-x$;   (c) $\dfrac{k+1}{3(k-1)}$;   (d) $\dfrac{x-4}{x+7}$;

(e) $-\dfrac{a-b}{3a+b}$; (f) $-\dfrac{x-3}{x+3}$

20. Since you may have forgotten some of what you once learned in arithmetic about mixed numbers, let us take a moment to review this subject. A mixed number consists of a whole number and a fraction. For example, $3\frac{1}{4}$, $7\frac{1}{2}$, $1\frac{3}{4}$, and $a + \dfrac{1}{b}$ are mixed numbers. In the first three expressions, notice that the plus sign is not used or needed for mixed numbers in arithmetic. However, it *is* required for mixed numbers in algebra, as shown in the fourth example. This is an important difference, because $3\frac{1}{4}$ means $3 + \frac{1}{4}$, but $a\frac{1}{b}$ means a times $\dfrac{1}{b}$.

A mixed number can be changed into a simple fraction by adding the whole number and the fraction. But to do so you must have a common denominator. If you are converting only one mixed number, the common denominator will be the denominator of the fractional part of the mixed number. Thus,

$$3\frac{2}{5} = 3 + \frac{2}{5} = \frac{3}{1} + \frac{2}{5} = \frac{15}{5} + \frac{2}{5} = \frac{17}{5}$$

Or, using an algebraic mixed number,

$$a + \frac{k}{y} = \frac{ay}{y} + \frac{k}{y} = \frac{ay + k}{y}$$

Similarly,

$$x - y - \frac{x^2}{x + y} = \frac{x - y}{1} - \frac{x^2}{x + y}$$
$$= \frac{(x - y)(x + y) - x^2}{x + y} = \frac{x^2 - y^2 - x^2}{x + y}$$
$$= \frac{y^2}{x + y}$$

As you develop more practice, you will find yourself performing some of the above steps mentally without putting them in writing.

Change the following mixed expressions into common fractions.

(a) $\quad 4\dfrac{7}{8} = $ _____

(b) $\quad 3 + \dfrac{4}{7} = $ _____

(c) $\quad \dfrac{a}{d} - 1 = $ _____

(d) $\quad 2k + \dfrac{3}{k} = $ _____

(e) $\quad 3a + b - \dfrac{2}{a} = $ _____

(f) $\quad a + c + d - \dfrac{1}{ac} = $ _____

(g) $\quad 2x^2 + 3x + 1 + \dfrac{x^2 + 1}{2x - 1} = $ _____

(h) $\quad a^2 + ab - b^2 - \dfrac{a^3 - 2b^3}{a - 2b} = $ _____

- - - - - - - - - - - - -

(a) $\dfrac{39}{8}$; (b) $\dfrac{25}{7}$; (c) $\dfrac{a - d}{d}$; (d) $\dfrac{2k^2 + 3}{k}$; (e) $\dfrac{3a^2 + ab - 2}{a}$;

(f) $\dfrac{a^2c + ac^2 + acd - 1}{ac}$; (g) $\dfrac{4x^3 + 5x^2 - x}{2x - 1}$ (you could factor x out of the numerator);

(h) $-\dfrac{a^2b + 3ab^2 - 4b^3}{a - 2b}$ (you could factor b out of the numerator, but it is not necessary)

21. Here is another example of the procedure for changing an algebraic fraction into a mixed number:

$$\frac{4x^2 + 8x - 3}{2x} = \frac{4x^2}{2x} + \frac{8x}{2x} - \frac{3}{2x} = 2x + 4 - \frac{3}{2x}$$

Follow this procedure to change the following fractions into mixed expressions.

(a) $\dfrac{3x^3 + x^2 - 4x + 7}{x} =$ _____

(b) $\dfrac{12x^2 + 9x + 2}{3x} =$ _____

(c) $\dfrac{4a^2b^2 - 6ab - 5}{2ab} =$ _____

(d) $\dfrac{15x^3 - 10x^2 + 5}{5x} =$ _____

- - - - - - - - - - - - - -

(a) $3x^2 + x - 4 + \dfrac{7}{x}$;   (b) $4x + 3 + \dfrac{2}{3x}$;   (c) $2ab - 3 - \dfrac{5}{2ab}$;

(d) $3x^2 - 2x + \dfrac{1}{x}$

22. Changing a fraction with a binomial denominator into a mixed expression is slightly more involved than, although similar to, long division in arithmetic.

Example: Change $\dfrac{x^2 + 3x + 4}{x + 1}$ into a mixed expression.

Solution: Dividing is done with the first term of the binomial divisor. Do not try to divide with the second term! It is simply along for the ride. First divide x^2 by x (the first term of the divisor). This gives x as the first term of the quotient (written above x^2).

$$\begin{array}{r} x \phantom{{}+ 3x + 4} \\ x + 1 \overline{\smash{\big)}\, x^2 + 3x + 4} \end{array}$$

Then multiply both terms of the divisor by the first term of the quotient x, and subtract the resulting product from the dividend. (*Remember*: Subtraction changes *all* signs.)

$$\begin{array}{r} x \phantom{{}+ 3x + 4} \\ x + 1 \overline{\smash{\big)}\, x^2 + 3x + 4} \\ \underline{x^2 + 1x} \phantom{{}+ 4} \\ 2x \phantom{{}+ 4} \end{array}$$

Continue (very much as in long division) until the division is complete.

$$\begin{array}{r} x + 2 \\ x + 1 \overline{\smash{\big)}\ x^2 + 3x + 4} \\ \underline{x^2 + 1x} \\ 2x + 4 \\ \underline{2x + 2} \\ + 2 \end{array} \qquad = x + 2 + \frac{2}{x + 1}$$

In the last term of the answer the remainder becomes the numerator and the divisor the denominator, just as in arithmetic.

Change the following fractions into mixed expressions.

(a) $\dfrac{2a^2 + 3a + 2}{a + 2}$
(c) $\dfrac{5y^2 - 7y + 1}{y - 1}$

(d) $\dfrac{4x^3 - x + 1}{2x - 1}$

(b) $\dfrac{k^3 - 1}{k - 1}$ (*Hint*: Leave space between k^3 and 1 for the missing powers of k^2 and k.)

- - - - - - - - - - - - - - - - -

(a)
$$\begin{array}{r} 2a - 1 \\ a + 2 \overline{\smash{\big)}\ 2a^2 + 3a + 2} \\ \underline{2a^2 + 4a} \\ - a + 2 \\ \underline{- a - 2} \\ + 4 \end{array} \qquad = 2a - 1 + \frac{4}{a + 2}$$

(b)
$$\begin{array}{r} k^2 - k - 1 \\ k - 1 \overline{\smash{\big)}\ k^3 \qquad\quad + 1} \\ \underline{k^3 - k^2} \\ - k^2 \\ \underline{- k^2 + k} \\ - k \\ \underline{- k + 1} \\ - 2 \end{array} \qquad = k^2 - k - 1 - \frac{2}{k - 1}$$

(c)
$$\begin{array}{r} 5y - 2 \\ y - 1 \overline{\smash{\big)}\ 5y^2 - 7y + 1} \\ \underline{5y^2 - 5y} \\ - 2y \\ \underline{- 2y + 2} \\ - 1 \end{array} \qquad = 5y - 2 - \frac{1}{y - 1}$$

(d)
$$\begin{array}{r} 2x^2 + x \\ 2x - 1 \overline{\smash{\big)}\ 4x^3 \qquad\quad - x + 1} \\ \underline{4x^3 - 2x^2} \\ + 2x^2 \\ \underline{2x^2 - x} \\ + 1 \end{array} \qquad = 2x^2 + x + \frac{1}{2x - 1}$$

23. A *complex fraction* is one that has one or more fractions in the numerator, the denominator, or both, such as those shown in review item 23.

Below are some further examples and practice problems for the two methods of changing complex fractions into simple fractions. The procedures are restated for your convenience.

Method 1:   Multiply both numerator and denominator by the LCD of all fractions appearing in them, and reduce as necessary.

Example 1:   Simplify $\dfrac{\frac{2}{3}}{\frac{3}{4}}$.

Solution: Since the LCD = 12 (3 · 4), we have

$$\frac{\frac{2}{3}(12)}{\frac{3}{4}(12)} = \frac{8}{9}$$

Example 2:   Simplify $\dfrac{\frac{x^2 - y^2}{4}}{\frac{x + y}{2}}$.

Solution: Since the LCD = 4, we have

$$\frac{\frac{x^2 - y^2}{4}}{\frac{x + y}{2}} = \frac{\left(\frac{x^2 - y^2}{4}\right)(4)}{\left(\frac{x + y}{2}\right)(4)} = \frac{x^2 - y^2}{2(x + y)} = \frac{(x + y)^{\,1}(x - y)}{2(x + y)_{\,1}} = \frac{x - y}{2}$$

Simplify the following.

(a) $\dfrac{\frac{7}{2}}{\frac{2}{3}} = $ _____

(b) $\dfrac{2\frac{2}{3}}{\frac{4}{5}} = $ _____

(c) $\dfrac{10\frac{1}{2}}{1 + \frac{1}{5}} = $ _____

(d) $\dfrac{1 - \frac{x}{y}}{1 - \frac{x^2}{y^2}} = $ _____

Method 2:   Combine the terms in the numerator and in the denominator, and then divide the numerator by the denominator.

Example 1: $\dfrac{\frac{2}{3} - \frac{1}{2}}{\frac{3}{4} - \frac{1}{2}} = \dfrac{\frac{4-3}{6}}{\frac{3-2}{4}} = \dfrac{\frac{1}{6}}{\frac{1}{4}} = \dfrac{1}{6} \div \dfrac{1}{4} = \dfrac{1}{6} \cdot \dfrac{4}{1} = \dfrac{2}{3}$

Example 2: $\dfrac{1 + \frac{2}{x}}{1 - \frac{4}{x^2}} = \dfrac{\frac{x+2}{x}}{\frac{x^2-4}{x^2}} = \dfrac{x+2}{x} \div \dfrac{x^2-4}{x^2}$

$$= \dfrac{(x+2)^1}{x_1} \cdot \dfrac{x^2}{(x-2)(x+2)_1} = \dfrac{x}{x-2}$$

Simplify the following.

- - - - - - - - - - - - -

(e) $\dfrac{\frac{1}{z} + 1}{\frac{1}{z^2} - 1} =$

(f) $\dfrac{1 - \frac{2}{a} - \frac{3}{a^2}}{1 + \frac{1}{a}} =$

(g) $\dfrac{m + \frac{m}{n}}{\frac{1}{n} + \frac{1}{n^2}} =$

(h) $\dfrac{t - \frac{25}{t}}{t + 5} =$

- - - - - - - - - - - -

(a) $\dfrac{\frac{7}{2}\,(6)}{\frac{2}{3}\,(6)} = \dfrac{21}{4} = 5\frac{1}{4};$ (b) $\dfrac{\frac{8}{3}\,(15)}{\frac{4}{5}\,(15)} = \dfrac{40}{12} = \dfrac{10}{3} = 3\frac{1}{3};$ (c) $\dfrac{\frac{21}{2}\,(10)}{\frac{6}{5}\,(10)} = \dfrac{105}{12} = 8\frac{3}{4};$

(d) $\dfrac{\left(\frac{y-x}{y}\right)(y^2)\,y}{\left(\frac{y^2-x^2}{y^2}\right)(y^2)_1} = \dfrac{y(y-x)}{y^2-x^2} = \dfrac{y(y-x)^1}{1(y-x)(y+x)} = \dfrac{y}{y+x};$

(e) $\dfrac{(1+z)^1}{z} \cdot \dfrac{z^2}{(1-z)(1+z)_1} = \dfrac{z}{1-z};$ (f) $\dfrac{(a+1)^1(a-3)}{a^2} \cdot \dfrac{a}{(a+1)_1} = \dfrac{a-3}{a};$

(g) $\dfrac{m(n+1)^1}{n} \cdot \dfrac{n^2}{(n+1)_1} = mn;$ (h) $\dfrac{(t-5)(t+5)^1}{t} \cdot \dfrac{1}{(t+5)_1} = \dfrac{t-5}{t};$

UNIT SIX

Exponents, Roots, and Radicals

Review Item	Ref Page	Example
1. A positive integral exponent indicates how many times the base is to be taken as a factor.	107	$x^5 = x \cdot x \cdot x \cdot x \cdot x$
2. When multiplying like bases, add their exponents; when dividing like bases, subtract the exponents.	108	$z^3 \cdot z^4 = z^{3+4} = z^7$ $x^5 \div x^2 = x^{5-2} = x^3$
3. To find the power of the power of a base, keep the base and multiply the exponents.	109	$(c^3)^4 = c^{12}$ $(a^b)^c = a^{bc}$
4. The positive exponent of a base appearing in the denominator of a fraction becomes a negative exponent when the base is moved to the numerator.	109	$\dfrac{1}{k^2} = k^{-2}; \ x^{-4} = \dfrac{1}{x^4}$ $\dfrac{2x^2y}{a^3b} = 2x^2ya^{-3}b^{-1}$
5. The root of a number is one of its equal factors.	110	$\sqrt{4} = 2$, since $2 \cdot 2 = 4$ $\sqrt[3]{8} = 2$, since $2 \cdot 2 \cdot 2 = 8$ $\sqrt[4]{16} = 2$, since $2 \cdot 2 \cdot 2 \cdot 2 = 16$

Review Item	Ref Page	Example
6. To extract the square root of any number, follow the procedure shown in reference item 6.	111	See reference item 6.
7. The square root of a fraction may be found by dividing the square root of the numerator by the square root of the denominator.	113	$\sqrt{\dfrac{x}{y}} = \dfrac{\sqrt{x}}{\sqrt{y}}$ $\sqrt{\dfrac{4}{9}} = \dfrac{\sqrt{4}}{\sqrt{9}} = \dfrac{2}{3}$
8. When the denominator of a fractional radicand is not a perfect square, multiply both the numerator and the denominator by a number that will make the denominator a perfect square. This is called *rationalizing the denominator.*	113	$\sqrt{\dfrac{1}{5}} = \sqrt{\dfrac{1 \cdot 5}{5 \cdot 5}} =$ $\sqrt{\dfrac{5}{25}} = \dfrac{\sqrt{5}}{5}$ or $\dfrac{1}{5}\sqrt{5}$
9. In a fractional exponent, the denominator is the index of a radical, and the numerator is the power of the radicand.	114	$x^{3/4} = 3\sqrt{x^2}$ $x^{1/2} = \sqrt{x}$ $x^{3/4} = 4\sqrt{x^3}$
10. The square root of a number is always considered to be its *absolute value.*	115	$\sqrt{x^2} = |x|$
11. Numbers that can be expressed as the ratio of two integers are called *rational numbers.* Numbers that cannot be so expressed are called *irrational numbers.*	115	$2 = \dfrac{2}{1}; \quad \dfrac{2 \cdot 3}{3 \cdot 4} = \dfrac{1}{2};$ $.6666 \ldots = \dfrac{2}{3}$ $\sqrt{3}; \ \pi$

Review Item	Ref Page	Example
12. A radical can be simplified if: • The radicand is a perfect square. • The radicand contains a factor that is a perfect square. • The radicand is a fraction.	116	$\sqrt{81} = 9$ $\sqrt{80} = \sqrt{16} \cdot \sqrt{5} = 4\sqrt{5}$ $\sqrt{\dfrac{1}{3}} = \sqrt{\dfrac{1 \cdot 3}{3 \cdot 3}} = \dfrac{\sqrt{3}}{3}$ or $\dfrac{1}{3}\sqrt{3}$
13. *Similar radicals* are radicals having the same index and the same radicand. They can be combined by addition and subtraction.	117	$2\sqrt{3} + 4\sqrt{3} - \sqrt{3} = 5\sqrt{3}$
14. In multiplying radicals, the product of the square roots of two or more nonnegative real numbers is equal to the square root of their product. The square of the square root of a positive number equals the number.	118	$\sqrt{a} \cdot \sqrt{b} = \sqrt{ab}$ $\sqrt{a} \cdot \sqrt{b} \cdot \sqrt{c} = \sqrt{abc}$ $\sqrt{2} \cdot \sqrt{5} \cdot 3 = \sqrt{30}$ $(\sqrt{a})^2 = a$ $(\sqrt{8})^2 = 8$
15. To multiply binomial radicals, follow the normal rules for multiplication together with the rule for multiplying monomial radicals.	119	$\sqrt{2}(\sqrt{2} - 3) = 2 - 3\sqrt{2};$ $(2 + \sqrt{3})(3 - \sqrt{3}) = 6 + \sqrt{3} - 3.$
16. To divide monomials that contain radicals, divide the coefficients and radicals separately; then simplify, if possible.	120	$\dfrac{8\sqrt{40}}{4\sqrt{5}} = \dfrac{8}{4}\sqrt{\dfrac{40}{5}} = 2\sqrt{8}$ $= 2 \cdot 2\sqrt{2} = 4\sqrt{2};$ $\dfrac{15\sqrt{x^7}}{3\sqrt{x^4}} = \dfrac{15}{3}\sqrt{\dfrac{x^7}{x^4}}$ $= 5\sqrt{x^3}$ $= 5\sqrt{x^2} \cdot x = 5x\sqrt{x}$

Review Item	Ref Page	Example
17. To rationalize a binomial denominator containing a radical, multiply the numerator and the denominator by the denominator with its middle sign changed.	120	Rationalize: $\dfrac{\sqrt{2}+3}{\sqrt{2}-1}$ $\dfrac{\sqrt{2}+3}{\sqrt{2}-1} \quad \dfrac{\sqrt{2}+1}{\sqrt{2}+1}$ $= \dfrac{2+4\sqrt{2}+3}{2-1} \quad = 4\sqrt{2}+5$

18. To solve a radical equation:
 	
 Rearrange the terms so that the radical is alone on one side of the equation.
 	
   • Square both sides.
 	
   • Solve for the unknown.
 	
   • Check your answer (the *root*) by substituting it back into the original equation.
	121	Solve: $3 + \sqrt{2x} = 7$
		
		 $\sqrt{2x} = 7 - 3 = 4$
		
		 $2x = 16$
		
		 $x = 8$
		
		 $3 + \sqrt{2} \cdot 8 = 7$
		 $3 + \sqrt{16} = 7$
		 $3 + 4 = 7$
		 $7 = 7$

UNIT SIX REFERENCES

1. Another way of defining the role of the exponent, which we covered first in review item 24, Unit 1, is as follows: If a is a real number and m is a positive integer,

$$a^m = \overbrace{a \cdot a \cdot a \cdot a \cdots a}^{m \text{ factors}}$$

Remember that a is called the *base,* m is called the *exponent,* and the result $(a \cdot a \cdot a \cdot a \cdots a)$ is called the *power.* The value of m tells how many times the base is to be taken as a factor. Thus, $7^2 = 7 \cdot 7 = 49$; $(ab^2)^2 = (ab^2)(ab^2) = a^2b^4$; $(-2x^2y^3)^2 = (-2x^2y^3)(-2x^2y^3) = 4x^4y^6$; and so on.

Use the rule of positive integral exponents to work out the following problems.

(a) $8^2 = $ _____   (f) $(-2a^2b^3)^2 = $ _____

(b) $(-3)^2 = $ _____   (g) $(xy^2z^3)^2 = $ _____

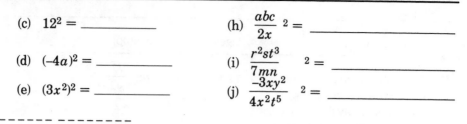

(c) $12^2 = $ _____

(d) $(-4a)^2 = $ _____

(e) $(3x^2)^2 = $ _____

(h) $\dfrac{abc}{2x}\,2 = $ _____

(i) $\dfrac{r^2st^3}{7mn}\,2 = $ _____

(j) $\dfrac{-3xy^2}{4x^2t^5}\,2 = $ _____

(a) 64;   (b) 9;   (c) 144;   (d) $16a^2$;   (e) $9x^4$;   (f) $4a^4b^6$;   (g) $x^2y^4z^6$;

(h) $\dfrac{a^2b^2c^2}{4x^2}$;   (i) $\dfrac{r^4s^2t^6}{49m^2n^2}$;   (j) $\dfrac{9x^2y^4}{16x^4t^{10}}$

2. You may remember these rules from Unit 3 (review items 13 and 18). We include them again here to make sure you have the necessary preparation for the work we will be doing in this unit. We can restate the rules in more general terms as follows:

$$a^m \cdot a^n = a^{m+n};\text{ and } a^m \div a^n = a^{m-n}, \text{ if } m \text{ and } n \text{ are positive integers.}$$

Thus, $b^4 \cdot b^2 = b^{4+2} = b^6$; and $b^4 \div b^2 = b^{4-2} = b^2$.

Applying the rule for division, what happens when we get the result $x^5 \div x^5 = x^{5-5} = x^0$? Let us see. We know that any number other than zero divided by itself equals one. Therefore, $x^5 \div x^5 = 1$. From these results and the substitution law of equality we therefore have the following:

$$\text{If } \frac{x^5}{x^5} = 1 \text{ and } \frac{x^5}{x^5} = x^0, \text{ then } x^0 = 1.$$

In more general terms:

$$a^0 = 1 \quad \text{if } a \neq 0.$$

Apply the above rules to the following problems. Assume that variables are not equal to zero.

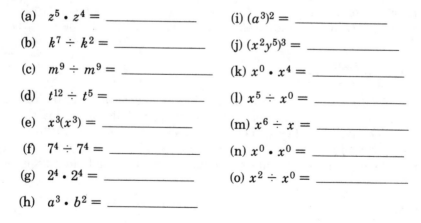

(a) $z^5 \cdot z^4 = $ _____

(b) $k^7 \div k^2 = $ _____

(c) $m^9 \div m^9 = $ _____

(d) $t^{12} \div t^5 = $ _____

(e) $x^3(x^3) = $ _____

(f) $7^4 \div 7^4 = $ _____

(g) $2^4 \cdot 2^4 = $ _____

(h) $a^3 \cdot b^2 = $ _____

(i) $(a^3)^2 = $ _____

(j) $(x^2y^5)^3 = $ _____

(k) $x^0 \cdot x^4 = $ _____

(l) $x^5 \div x^0 = $ _____

(m) $x^6 \div x = $ _____

(n) $x^0 \cdot x^0 = $ _____

(o) $x^2 \div x^0 = $ _____

------- ---------

(a) $z^{5+4} = z^9$; (b) $k^{7-2} = k^5$; (c) $m^{9-9} = m^0 = 1$; (d) $t^{12-5} = t^7$; (e) $x^{3+3} = x^6$; (f) $7^{4-4} = 7^0 = 1$; (g) $2^{4+4} = 2^8$; (h) $a^3 \cdot b^2 = a^3 b^2$ (cannot be simplified because no common base); (i) $a^{3 \cdot 2} = a^6$; (j) $x^6 y^{15}$; (k) x^4; (l) x^5; (m) x^5; (n) 1; (o) x^2

3. This rule should be familiar from review item 14, Unit 3. If not, review the item and the corresponding reference item; then apply the rule to the following problems.

(a) $(x^4)^2 = $ _____

(b) $(a^2 b^3)^3 = $ _____

(c) $\left(\dfrac{a}{b}\right)^3 = $ _____

(d) $(t^m)^n = $ _____

(e) $(c^3)^2 c^2 = $ _____

(f) $\left(\dfrac{x^2 y}{xg^3}\right)^3 = $ _____

(g) $g \cdot (g^2)^2 = $ _____

(h) $\left(\dfrac{3a^3 b^2}{4mnz}\right)^3 = $ _____

------- ---------

(a) x^8; (b) $a^6 b^9$; (c) $\dfrac{a^3}{b^3}$; (d) t^{mn}; (e) $c^{3 \cdot 2}(c^2) = c^6 \cdot c^2 = c^8$; (f) $\dfrac{x^6 y^3}{x^3 g^9}$;

(g) $g^2 g^{22} = g^6$; (h) $\dfrac{27a^9 b^6}{64m^3 n^3 z^3}$

4. So far, we have been working with exponents that are positive numbers. Now we need to consider the possibility of negative exponents. In review items 18 and 19, Unit 3, we mentioned that a fraction such as $\dfrac{1}{x^4}$ could also be written as x^{-4}. This is justified by the rule for the division of like bases stated in item 2 above: $a^m \div a^n = a^{m-n}$.

Example: $x^2 \div x^5 = \dfrac{x \cdot x}{x \cdot x \cdot x \cdot x \cdot x} = \dfrac{x^2}{x^5} = \dfrac{1}{x^3}$

Applying our rule for division, we have:

$$\dfrac{x^2}{x^5} = x^{2-5} = x^{-3}$$

Generalizing from this example, we can say:

If $a \neq 0$ and if n is a positive integer, then $a^{-1} = \dfrac{1}{a}$, and a^{-n} $= \dfrac{1}{a^n}$. Conversely, $\dfrac{1}{a^{-n}} = a^n$.

Use this definition in working the following problems. Write your answers using positive exponents unless otherwise indicated.

(a)   Write x^{-3} using a positive exponent. _____

(b)   Write $\dfrac{1}{b^4}$ using a negative exponent. _____

(c)   Would it be consistent with our definition of a reciprocal to say that a^{-1} is the reciprocal of a? _____

(d)   Write the answer to $b^3 \div b^7$ using positive exponents. _____

(e)   Write the answer to $b^3 \div b^7$ using negative exponents. _____

(f)   Write $\dfrac{1}{x^{-4}}$ using a positive exponent. _____

(g) $x^5 \div x^3 =$ _____; or, using negative exponents, _____

(h) $x^3 \div x^5 =$ _____; or, using negative exponents, _____

(i) $x^{12} \div x^{13} =$ _____; or, using negative exponents, _____

(j) $x^6 \div x^2 =$ _____; or, using negative exponents, _____

- - - - - - - - - - - - - -

(a) $\dfrac{1}{x^3}$;   (b) b^{-4};   (c) yes, because the reciprocal of a is $\dfrac{1}{a}$;   (d) $\dfrac{1}{b^4}$;   (e) b^{-4};

(f) x^4;   (g) x^2, $\dfrac{1}{x^{-2}}$;   (h) $\dfrac{1}{x^2}$, x^{-2};   (i) $\dfrac{1}{x}$, x^{-1};   (j) x^4, $\dfrac{1}{x^{-4}}$

5.   The inverse of raising a number to a power is *extracting the root* of a number. We touched on this briefly in Unit 4. Raising a to the power m is indicated by the expression a^m. Similarly, extracting the mth root of a is indicated by $\sqrt[m]{a}$. The symbol $\sqrt{}$ is called the *radical sign*, a is termed the *radicand*, and m is known as the *index*. If no index is written, it is assumed to be 2. Thus, $\sqrt{a}$ means the positive square root of a. Hence the following definition.

$$n = \sqrt[m]{a}, \text{ if } n^m = a$$

In this case, n is called the mth root of a.

Examples:

$\sqrt{9}\ \ = 3$ because $3^2 = 9$

$\sqrt{25} = 5$ because $5^2 = 25$

$\sqrt[3]{8}\ \ = -2$ because $2^3 = 8$

$\sqrt[3]{-8} = -2$ because $(-2)^3 = -8$

$\sqrt[4]{81} = 3$ because $3^4 = 81$

So we can arrive at this definition:

> *A number is an nth root of a given number if it is one of the n equal factors of that number.*

Examples: A square root of a (indicated by $\sqrt{a}$) is one of two equal factors of a.
A cube root of a (indicated by $\sqrt[3]{a}$) is one of three equal factors of a.
A fourth root of a (indicated by $\sqrt[4]{a}$) is one of four equal factors of a.

Many numbers have roots of some kind that are exact integers or non-integral rational numbers. The following problems involve numbers of this type. Find the roots. If the index is an even number, give the absolute value.

(a) $\sqrt{49} = $ _____   (f) $\sqrt{9} = $ _____

(b) $\sqrt{64} = $ _____   (g) $\sqrt{81}$ _____

(c) $\sqrt[3]{27} = $ _____   (h) $\sqrt[5]{32} = $ _____

(d) $\sqrt[4]{16} = $ _____   (i) $\sqrt[3]{a^6} = $ _____

(e) $\sqrt[3]{125} = $ _____   (j) $\sqrt[5]{a^{25}} = $ _____

------ ----------

(a) 7; (b) 8; (c) 3; (d) 2; (e) 5; (f) 3; (g) 9; (h) 2; (i) a^2; (j) a^5

6. As stated earlier, the roots of many numbers can be written as integers or as non-integral rational numbers. We can often find these values by inspection. The real problem arises when the radicand is too large to solve by inspection or does not give an integral root.

It would be difficult and time consuming to use trial and error to find the square root of a radicand as large as 106,929 which happens to be a perfect square. To solve such problems without the aid of a calculator or a square root table, we need a practical and consistent method. With a little practice, the method shown below will be quite easy.

Extracting the Square Root

The six-digit number 106,929 is a perfect square. To find its square root we proceed as follows:

(1) Separate the radicand into groups of two figures each, counting from the decimal point (from right to left in this case).

$$10\ 69\ 29.$$

(2) Under the first pair, 10, place 9, the largest perfect square less than 10. Above the first pair place 3, the square root of 9. Subtract 9 from 10 to obtain the remainder 1.

$$\begin{array}{r} 3 \\ \sqrt{10\ 69\ 29.} \\ \underline{9} \\ 1 \end{array}$$

(3) Bring down 69, the next pair of figures. The new dividend is 169.

$$\begin{array}{r} 3 \\ \sqrt{10\ 69\ 29.} \end{array}$$

(4) To get the new divisor, double 3 (the first figure in the root) and get 6 as the first figure of the divisor.

$$\begin{array}{r} 9 \\ 6\ \overline{|\ 1\ 69} \end{array}$$

(5) There is one more figure in the divisor, and it is also the next figure in the root. To find it, divide 6 into 16 (the first two figures of the remainder). Since it will go twice, place a 2 beside the 6 to get 62, the complete divisor, and also place the 2 above the 69, as the next figure of the root.

$$\begin{array}{r} 3\ 2 \\ \sqrt{10\ 69\ 29.} \\ 9 \\ 62\ \overline{|\ 1\ 69} \end{array}$$

(6) Now multiply this 2, the last digit of the partial root, times 62 and subtract the product, 124, from 169 to get the new remainder of 45.

$$\begin{array}{r} 3\ 2\ 7 \\ \sqrt{10\ 69\ 29.} \\ 9 \end{array}$$

(7) Bring down 29, the last pair of figures in the original number. The new dividend is now 4529. Again double the partial root (32) to get 64 and place this in front of 4529. By trial we find that 64 will go into 452 (the first three figures of the remainder) seven times. Therefore, place a 7 after 64 and also above 29 as the last figure in the root.

$$\begin{array}{r} 62\ \overline{|\ 1\ 69\ 29} \\ \underline{1\ 24} \\ 647 \quad 45\ 29 \end{array}$$

(8) Multiply 647 by 7 and place the product under 4529. Since there is no remainder, the square root of 106,929 is the whole number 327. Had there been a remainder you would place a decimal point after the final 9 (in 10 69 29) and start bringing down pairs of zeros to find the value of the root to as many decimal places as you required.

$$\begin{array}{r} 3\ 2\ 7 \\ \sqrt{10\ 69\ 29.} \\ 9 \\ 62\ \overline{|\ 1\ 69\ 29} \\ \underline{1\ 24} \\ 647\ \overline{|\ 45\ 29} \\ \underline{45\ 29} \end{array}$$

7. So far, we have only extracted roots of integers. Extracting the roots of a fraction is easy if the root is a rational number. In this case, the root is found by extracting the root of the numerator and dividing it by the root of the denominator. Putting this into symbols, we can write the following definition:

$$\sqrt[n]{\frac{a}{b}} = \frac{\sqrt[n]{a}}{\sqrt[n]{b}}$$

Examples: $\sqrt{\frac{9}{25}} = \frac{\sqrt{9}}{\sqrt{25}} = \frac{3}{5}; \quad \sqrt{\frac{25x^2}{36y^4}} = \sqrt{\frac{25x^2}{36y^4}} = \frac{5x}{6y^2}$

Work the following problems, using the definition given above to guide you. Assume that all variables are positive.

(a) $\sqrt{\frac{1}{9}} = $ _____

(b) $\sqrt{\frac{16}{49}} = $ _____

(c) $\sqrt{\frac{4}{64}} = $ _____

(d) $\sqrt{\frac{x^4}{25}} = $ _____

(e) $\sqrt{\frac{9t^2}{100s^8}} = $ _____

(f) $\sqrt{\frac{m^4n^8}{81p^2}} = $ _____

(g) $\sqrt{\frac{49x^2y^4}{144b^6}} = $ _____

(h) $\sqrt{\frac{81a^4}{121b^4}} = $ _____

- - - - - - - - - - - - -

(a) $\frac{1}{3}$; (b) $\frac{4}{7}$; (c) $\frac{1}{4}$; (d) $\frac{x^2}{5}$; (e) $\frac{3t}{10s^4}$; (f) $\frac{m^2n^4}{9p}$; (g) $\frac{7xy^2}{12b^3}$;

(h) $\frac{9a^2}{11b^2}$

8. Our rule for finding the square root of a fraction works very well if the numerator and denominator are both perfect squares. But what if they are not? Then we apply the rule given in review item 8: We multiply both numerator and denominator by a number that will make the denominator a perfect square.

Example: Find the approximate square root of the fraction $\frac{2}{3}$. In other words, $\sqrt{\frac{2}{3}} = ?$

Solution: Following our rule, we multiply the numerator and denominator by 3, giving us $\sqrt{\frac{6}{9}}$. From this we get $\frac{\sqrt{6}}{3}$, or, using the approximate root of 6*, $\frac{2.449}{3} = .816$. We could have taken the square root of 2 and divided it by the square root of 3 to begin with, but the method we used is faster and offers less chance for error.

Try the method yourself by finding the approximate square roots (to two decimal places) of the following fractions.

*See the Table of Powers and Roots at the back of this book for approximate values.

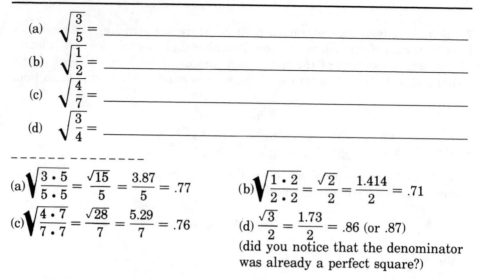

(a) $\sqrt{\dfrac{3}{5}} = $ _____

(b) $\sqrt{\dfrac{1}{2}} = $ _____

(c) $\sqrt{\dfrac{4}{7}} = $ _____

(d) $\sqrt{\dfrac{3}{4}} = $ _____

- - - - - - - - - - - - - -

(a) $\sqrt{\dfrac{3 \cdot 5}{5 \cdot 5}} = \dfrac{\sqrt{15}}{5} = \dfrac{3.87}{5} = .77$

(c) $\sqrt{\dfrac{4 \cdot 7}{7 \cdot 7}} = \dfrac{\sqrt{28}}{7} = \dfrac{5.29}{7} = .76$

(b) $\sqrt{\dfrac{1 \cdot 2}{2 \cdot 2}} = \dfrac{\sqrt{2}}{2} = \dfrac{1.414}{2} = .71$

(d) $\dfrac{\sqrt{3}}{2} = \dfrac{1.73}{2} = .86$ (or .87)

(did you notice that the denominator was already a perfect square?)

9.   The rules for handling exponents we have studied so far relate strictly to integral exponents. Now we need to consider fractional exponents. Reflecting upon the possible meaning of a fractional exponent such as $x^{1/2}$ (where x is a positive number), we might get some clue by observing that $x^{1/2} \cdot x^{1/2} = x^{1/2 + 1/2} = x$, by the law of multiplication of terms having the same base. Since $\sqrt{x} \cdot \sqrt{x} = x$, and $x^{1/2} \cdot x^{1/2} = x$, we might suspect that $x^{1/2} = \sqrt{x}$. Also, since the product of $x^{1/3} \cdot x^{1/3} \cdot x^{1/3}$ is $x^{(1/3) + (1/3) + (1/3)}$, which is x, it would seem consistent to say that $x^{1/3} = \sqrt[3]{x}$.

Let us consider the quantity x^n. If we want the value of $x^{1/n} \cdot x^{1/n} \cdot x^{1/n} \ldots x^{1/n}$ (for n factors), we could write this as $(x^{1/n})^n$, and our rule for finding the power of the power of a base (review item 3 above) would allow us to write this as $x^{(1/n)(n)} = x$. Although this certainly is not a proof, it seems clear that since $(x^{1/n})^n = x$ and $(\sqrt[n]{x})^n = x$, we can say that $x^{1/n} = \sqrt[n]{x}$. If we accept the above argument, then it would be consistent to say that $x^{m/n} = x^{(1/n)(m)} = (\sqrt[n]{x})^m = \sqrt[n]{x^m}$. It appears, therefore, that in a fractional exponent, the denominator is the index of a radical, and the numerator is the power of the radicand. Thus, $x^{2/3} = \sqrt[3]{x^2}$; $x^{1/2} = \sqrt{x}$; $x^{3/4} = \sqrt[4]{x^3}$.

With the foregoing in mind, write the answers to the following problems.

(a) $4^{1/2} = $ _____

(b) $25^{1/2} = $ _____

(c) $\left(x^{1/2}\right)^2 = $ _____

(d) $8^{1/3} = $ _____

(e) $27^{1/3} = $ _____

(f) $(-8)^{1/3} = $ _____

(g) $(16a^8)^{1/2} = $ _____ $(a > 0)$

(h) $\left(\dfrac{4}{9}\right)^{1/2} = $ _____

- - - - - - - - - - - - - -

(a) 2;   (b) 5;   (c) x;   (d) 2;   (e) 3;   (f) –2;   (g) $4a^4$;   (h) $\dfrac{2}{3}$

10. Having looked into the subject of fractional exponents in review item 9, we can now examine exactly what $\sqrt{x^2}$ means.

Suppose we wish to find the value of $\sqrt{(-2)^2}$. We could write this with fractional exponents as $\left[(-2)^2\right]^{1/2}$. On the basis of our prior discussions there are two ways to approach this problem.

Method 1: $\left[(-2)^2\right]^{1/2} = \left[(-2)(-2)\right]^{1/2} = 4^{1/2} = 2$

Method 2: $\left[(-2)^2\right]^{1/2} = (-2)^{2 \cdot 1/2} = (-2)^1 = -2$

Since we cannot have two different answers to the same problem, we must agree on a method that will yield a unique (that is, one) answer. Accordingly, we establish the following definition:

$$\sqrt{x^2} = |x| \text{ or } (x^2)^{1/2} = |x| \text{ (the absolute value of } x)$$

If we follow this definition, we will always achieve a unique answer. Discover this for yourself in the following problems. Assume that the variables may be any real number.

(a) $\sqrt{(-3)^2} =$ _____

(b) $\sqrt{5^2} =$ _____

(c) $\sqrt{a^2} =$ _____

(d) $\sqrt{(-5)^2} =$ _____

(e) $\sqrt{m^2} =$ _____

(f) $\sqrt{(-7)^2} =$ _____

- - - - - - - - - - - - - -

(a) $|-3| = 3$;   (b) $|5| = 5$;   (c) $|a|$;   (d) $|-5| = 5$;   (e) $|m|$;   (f) $|-7| = 7$

11. In review items 4 and 5, Unit 2, we touched on the subject of rational and irrational numbers. What we say here will serve as a partial review of those items and also extend some of the ideas in this unit. A *rational number* is one that can be expressed as the ratio of two integers. The following are rational numbers:

(1) All integers (that is, all positive and negative whole numbers and zero)

$$3 = \frac{3}{1}$$

(2) Fractions whose numerator and denominator are already integers or become integers after simplification

$$\frac{2 \cdot 5}{4 \cdot 5} = \frac{1}{2}$$

(3) Decimals that "terminate" after a finite number of steps (i.e., all later terms in the expansion are zero)

$$\frac{687}{100} = 6.87000 \ldots$$

(4) Decimals that do not terminate but whose digits repeat themselves $0.3333333 \ldots \frac{1}{3}$

(5) Radical expressions whose radicands are perfect powers $\sqrt{36} = 6; \quad \sqrt[3]{8} = 2$

Numbers that cannot be expressed as the ratio of two integers are known as *irrational numbers.* The square root of 2 is an example of an irrational number, because it cannot be changed into a whole number or a simple fraction. Its exact value, $\sqrt{2}$, cannot be written as a rational number. An approximation can be made, but this approximation will be a nonterminating, nonrepeating decimal.

Check the numbers that are rational.

_____(a) $\sqrt{4}$ _____ (e) $\sqrt{81}$

_____(b) $\sqrt{3}$ _____(f) 13

_____(c) 1.748   _____(g) .66666 $\cdots$

_____(d) $\dfrac{1.5}{2}$ _____(h) $\sqrt{8}$

- - - - - - - - - - - - - - -

You should have checked the following. (The numbers given in parentheses refer to the five types of rational numbers given above.)

(a) $\sqrt{4}$ (#5);   (c) 1.748 (#3);   (d) $\dfrac{1.5}{2} = \dfrac{3}{4}$ (#2);   (e) $\sqrt{81}$ (#5);   (f) 13 (#1); (g) .66666 $\cdots$ (#4)

$\sqrt{3}$ and $\sqrt{8}$ are *irrational* numbers because they cannot be expressed as the ratio of two integers.

12.   It is often necessary to simplify radicals, but not all radicals can be simplified. Let us see how such simplifying is done.

Can $\sqrt{36}$ be simplified? Yes, by taking its square root. Thus, $\sqrt{36} = 6$.

Can $\sqrt{48}$ be simplified? Yes, if you recognize that it contains a perfect square (16) as a factor. Thus, $\sqrt{48} = \sqrt{16} \cdot \sqrt{3}$, or $4\sqrt{3}$.

Can $\sqrt{\dfrac{2}{3}}$ be simplified? Yes, merely by multiplying the numerator and denominator by 3, which will make the denominator a perfect square (as explained in review item 8.) This gives us $\sqrt{\dfrac{6}{9}}$ or $\dfrac{\sqrt{6}}{3}$ or $\dfrac{1}{3}\sqrt{6}$. (Remember that $\dfrac{x}{3}$ and $\dfrac{1}{3}x$ represent the same value.)

In summary, then, a radical can be simplified if:

- The radicand is a perfect square.
- The radicand contains a factor that is a perfect square.
- The radicand is a fraction.

Simplify the following radicals. All variables are positive.

(a) $\sqrt{81} =$ _____

(b) $\sqrt{32} =$ _____

(c) $\sqrt{50} =$ _____

(d) $\sqrt{49x^6} =$ _____

(e) $\sqrt{\dfrac{1}{4}} =$ _____

(f) $\sqrt{\dfrac{x^2}{y^2}} =$ _____

(g) $\sqrt{\dfrac{2}{7}} =$ _____

(h) $\sqrt{b^3} =$ _____ (*Hint:* $b^3 = b^2 \cdot b$)

(i) $\sqrt{\dfrac{1}{a}} =$ _____

(j) $4\sqrt{27} =$ _____

(k) $\sqrt{121} =$ _____

(l) $3\sqrt{128}$_____

- - - - - - - - - - - -

(a) 9;   (b) $4\sqrt{2}$;   (c) $5\sqrt{2}$;   (d) $7x^3$;   (e) $\dfrac{1}{2}$;   (f) $\dfrac{x}{y}$;   (g) $\dfrac{1}{7}\sqrt{14}$;   (h) $b\sqrt{b}$;
(i) $\dfrac{1}{a}\sqrt{a}$;   (j) $12\sqrt{3}$;   (k) 11;   (l) $24\sqrt{2}$

13.   Another useful operation is the combining of terms containing *similar radicals,* that is, radicals having the same index and the same radicand. Similar radicals may be combined by addition and subtraction, through application of the distributive law, as shown below:

$$a\sqrt{x} + b\sqrt{x} = (a + b)\sqrt{x}$$

Notice that $a\sqrt{x}$ and $b\sqrt{x}$ are like terms.

Examples: $4\sqrt{5} + 3\sqrt{5} - 2\sqrt{5} = (4 + 3 - 2)\sqrt{5} = 5\sqrt{5}$
$$3\sqrt{6} + 2\sqrt{5} - \sqrt{6} + 3\sqrt{5} = 3\sqrt{6} - \sqrt{6} + 2\sqrt{5} + 3\sqrt{5}$$
$$= (3 - 1)\sqrt{6} + (2 + 3)\sqrt{5}$$
$$= 2\sqrt{6} + 5\sqrt{5}$$

The similarity of the radicands may not always be so obvious when the radicals are not in simplified form.

Example: Combine $3\sqrt{8} - \sqrt{50} + 6\sqrt{32}.$

Solution: Simplifying the first term gives us:

$$3\sqrt{8} = 3\sqrt{4} \cdot 2 = 3 \cdot 2\sqrt{2} = 6\sqrt{2}$$

Simplifying the second term gives us:

$$\sqrt{50} = \sqrt{25} \cdot 2 = 5\sqrt{2}$$

Finally, the third term becomes:

$$6\sqrt{32} = 6\sqrt{16} \cdot 2 = 6 \cdot 4\sqrt{2} = 24\sqrt{2}$$

With the radicals in simplest form, we have three similar radicals which can be combined to give us:

$$6\sqrt{2} - 5\sqrt{2} + 24\sqrt{2} = 25\sqrt{2}$$

Use this procedure to simplify and combine the following.

(a) $\sqrt{8} - 2\sqrt{8} + 7\sqrt{8} =$ _____

(b) $7\sqrt{7} + 4\sqrt{7} - 3\sqrt{7} =$ _____

(c) $\sqrt{2} + 2\sqrt{3} + 3\sqrt{2} - \sqrt{3} =$ _____

(d) $2 + 6\sqrt{7} + 5 - 2\sqrt{7} =$ _____

(e) $4\sqrt{a} + 2\sqrt{b} - \sqrt{a} - \sqrt{b} =$ _____

(f) $4\sqrt{7} - \sqrt{28} - \sqrt{63} =$ _____

(g) $\sqrt{75} + 4\sqrt{3} + \sqrt{18} =$ _____

(h) $2\sqrt{9y} - \sqrt{4x} + 7\sqrt{x} - 3\sqrt{y} =$ _____

(i) $6\sqrt{b} + \sqrt{25a} - \sqrt{b} - 2\sqrt{a} =$ _____

(j) $\sqrt{2} + 3\sqrt{27} + 2\sqrt{50} - 4\sqrt{3} =$ _____

(a) $12\sqrt{2}$;   (b) $8\sqrt{7}$;   (c) $4\sqrt{2} + \sqrt{3}$;   (d) $7 + 4\sqrt{7}$;   (e) $3\sqrt{a} + \sqrt{b}$;   (f) $-\sqrt{7}$;
(g) $9\sqrt{3} + 3\sqrt{2}$;   (h) $5\sqrt{x} + 3\sqrt{y}$;   (i) $3\sqrt{a} + 5\sqrt{b}$;   (j) $11\sqrt{2} + 5\sqrt{3}$

14.   We have experimented with combining (that is, adding and subtracting) radicals. Now let us consider the procedures for multiplying them. Review item 14 presents two useful rules. Use them to guide you in working the following problems. This will be the easiest way to understand their application.

(a) $\sqrt{3} \cdot \sqrt{4} =$ _____   (f) $5\sqrt{2} \cdot 3\sqrt{2} =$ _____

(b) $\sqrt{5} \cdot \sqrt{7} =$ _____   (g) $\sqrt{13} \cdot \sqrt{13} =$ _____

(c) $\sqrt{2} \cdot \sqrt{18} =$ _____   (h) $\sqrt{12} \cdot \sqrt{3} =$ _____

(d) $\sqrt{6} \cdot \sqrt{6} =$ _____   (i) $3\sqrt{6} \cdot 5\sqrt{3} =$ _____

(e) $(\sqrt{9})^2 =$ _____   (j) $\sqrt{3} \cdot 2\sqrt{4} \cdot 3\sqrt{3} =$ _____

- - - - - - - - - - - - - -

(a) $\sqrt{12} = 2\sqrt{3}$; (b) $\sqrt{35}$; (c) $\sqrt{36} = 6$; (d) 6; (e) 9; (f) $15\sqrt{4} = 15 \cdot 2 = 30$;
(g) 13; (h) 6; (i) $15\sqrt{18} = 15 \cdot 3\sqrt{2} = 45\sqrt{2}$; (j) $6\sqrt{36} = 6 \cdot 6 = 36$

15. So far we have multiplied monomials only. Review item 15 illustrates the method for multiplying polynominals containing radicals. Here are two more examples.

Example 1:   A monomial times a binomial.

Solution: $\sqrt{3}(7 - \sqrt{3}) = 7\sqrt{3} - \sqrt{3} \cdot \sqrt{3} = 7\sqrt{3} - 3$

Here we used the distributive law as well as the rules for radicals.

Example 2:   A binomial times a binomial.

Solution: $(4 + \sqrt{2})(2 - \sqrt{2}) = 8 + 2\sqrt{2} - 4\sqrt{2} - \sqrt{4}$
$= 8 - 2\sqrt{2} - 2$
$= 6 - 2\sqrt{2}$

Here we used the familiar routine for finding the product of two binomials together with rules for radicals.

With these examples as a guide, perform the operations indicated below.

(a) $2(\sqrt{3} + \sqrt{7}) =$ _____

(b) $3(\sqrt{3} - 3) =$ _____

(c) $2\sqrt{3}(2 - \sqrt{6}) =$ _____

(d) $3\sqrt{5}(\sqrt{2} + 3) =$ _____

(e) $\sqrt{7}(2\sqrt{7} - \sqrt{25}) =$ _____

(f) $(\sqrt{2} + \sqrt{3})(\sqrt{2} - \sqrt{3}) =$ _____

(g) $(3 + \sqrt{5})(4 - 2\sqrt{5}) =$ _____

(h) $(\sqrt{x} + \sqrt{y})^2 =$ _____

(i) $(2\sqrt{3} - 1)^2 =$ _____

(j) $(a\sqrt{a} + b\sqrt{b})^2 =$ _____

------ ----------

(a) $2\sqrt{3} + 2\sqrt{7}$;   b) $3\sqrt{2} - 9$;   (c) $4\sqrt{3} - 6\sqrt{2}$;   (d) $3\sqrt{10} + 9\sqrt{5}$;   (e) $14 - 5\sqrt{7}$;
(f) $2 + \sqrt{6} - \sqrt{6} - 3 = -1$;   (g) $12 - 2\sqrt{5} - 10 = 2 - 2\sqrt{5}$;   (h) $x + 2\sqrt{xy} + y$;   (i)
$13 - 4\sqrt{3}$;   (j) $a^3 + 2ab\sqrt{ab} + b^3$

16.   Here are two more examples of the procedure for dividing radicals, with the procedure shown step by step.

	Example 1	*Example 2*
	$\dfrac{6\sqrt{24}}{2\sqrt{3}}$	$\dfrac{12\sqrt{a^6}}{3\sqrt{a^3}}$
(1) Divide the coefficients:	$3 \cdot \sqrt{\dfrac{24}{3}}$	$4 \cdot \sqrt{\dfrac{a^6}{a^3}}$
(2) Divide the radicals:	$3\sqrt{8}$	$4\sqrt{a^3}$
(3) Simplify, if possible:	$3(2\sqrt{2})$	$4(\sqrt{a^2} \cdot \sqrt{a})$
(4) Answers:	$6\sqrt{2}$	$4a\sqrt{a}$

Follow this general procedure in performing the following divisions. Use absolute values of square root terms.

(a) $\dfrac{6\sqrt{9}}{2\sqrt{3}} = \underline{\qquad}$

(b) $\dfrac{12\sqrt{8}}{6\sqrt{2}} = \underline{\qquad}$

(c) $\dfrac{15\sqrt{48}}{5\sqrt{6}} = \underline{\qquad}$

(d) $\dfrac{14\sqrt{a^5}}{7\sqrt{a^3}} = \underline{\qquad}$

(e) $\dfrac{6x^2\sqrt{10a}}{3x\sqrt{5a}} = \underline{\qquad}$

------ ----------

(a) $3\sqrt{3}$;   (b) $2\sqrt{4} = 4$;   (c) $3\sqrt{8} = 6\sqrt{2}$;   (d) $2\sqrt{a^2} = 2a$;   (e) $2x\sqrt{2}$

17.   A radical is not in its simplest form if the radicand is a fraction; a fraction is not in its simplest form if the denominator contains a radical. In review item 8, we showed that a radical in the denominator can be eliminated by multiplying the denominator and the numerator by a number that will make the denominator a perfect square. We called this rationalizing the denominator. Now we need to apply this procedure to binomial denominators.

Example: Rationalize the denominator of $\dfrac{3 + \sqrt{2}}{2 - \sqrt{2}}$

Solution: To make the radical drop out of the denominator, multiply it by another binomial whose terms are identical but whose middle sign is opposite, that is $(2 + \sqrt{2})$. This is known as the *conjugate binomial* of the denominator. The multiplication gives us:

$$\frac{(3 + \sqrt{2})}{(2 - \sqrt{2})} \cdot \frac{(2 + \sqrt{2})}{(2 + \sqrt{2})} = \frac{6 + 5\sqrt{2} + 2}{4 - 2} = \frac{8 + 5\sqrt{2}}{2} = 4 + \frac{5\sqrt{2}}{2}$$

This procedure is an application of a principle we covered in review item 14, Unit 4: $(a + b)(a - b) = a^2 - b^2$.

Use this procedure whenever possible. It is a valuable tool and will work with any binomial involving square roots. You will have a chance to apply it in rationalizing the denominators of the fractions below.

(a) $\dfrac{2}{\sqrt{2} + 3} =$ _____

(b) $\dfrac{3}{\sqrt{2} - 1} =$ _____

(c) $\dfrac{1 - \sqrt{3}}{2 + \sqrt{3}} =$ _____

(d) $\dfrac{3\sqrt{a} - 1}{\sqrt{a} + 1} =$ _____

- - - - - - - - - - - -

(a) $\dfrac{2(\sqrt{2} - 3)}{(\sqrt{2} + 3)(\sqrt{2} - 3)} = \dfrac{2\sqrt{2} - 6}{2 - 9} = \dfrac{2\sqrt{2} - 6}{-7}$ or $-\dfrac{2\sqrt{2} - 6}{7}$;

(b) $\dfrac{3(\sqrt{2} + 1)}{(\sqrt{2} - 1)(\sqrt{2} + 1)} = \dfrac{3\sqrt{2} + 3}{2 - 1} = 3\sqrt{2} + 3$;

(c) $\dfrac{(1 - \sqrt{3})(2 - \sqrt{3})}{(2 + \sqrt{3})(2 - \sqrt{3})} = \dfrac{2 - 3\sqrt{3} + 3}{4 - 3} = 5 - 3\sqrt{3}$;

(d) $\dfrac{(3\sqrt{a} - 1)(\sqrt{a} - 1)}{(\sqrt{a} + 1)(\sqrt{a} - 1)} = \dfrac{3a - 4\sqrt{a} + 1}{a - 1}$

18. Since we will be discussing equations generally in the next unit, here we will cover only the procedure for handling an equation that contains a radical.

Solving an equation means finding the value(s) of the variable (letter) that will make the equation a true statement. Thus, in the equation $3a = 6$, the solution is 2, because 2 times 3 is 6. The *root* of the equation—in this case, 2 —is the value which "satisfies" the equation, that is, reduces it to an identity.

A *radical equation* is simply an equation in which one letter appears as a radicand. Thus, $\sqrt{a} = 4$ and $\sqrt{3x} - 2 = 0$ are radical equations. To solve this type of equation, we proceed as shown in review item 18 and the example below.

Example: Solve $\sqrt{2x} - 2 = 4$.

Solution: (1) Rearrange the terms so that the radical is alone on one side of the equation: $\sqrt{2x} = 4 + 2$

(2) Square both sides: $2x = 36$

(3) Solve for the value of the unknown (dividing both sides by 2): $x = 18$

(3) Check by substituting the root

(18) back into the original equation: $\sqrt{2 \cdot 18} - 2 = 4$

$$6 - 2 = 4$$

$$4 = 4$$

Use this procedure to solve the following radical equations.

(a) $\sqrt{x} = 2$ (e) $2\sqrt{a + 2} = 6$

(b) $\sqrt{x + 4} = 3$ (f) $\sqrt{2x + 6} - 6 = 0$

(c) $\sqrt{x} = 1$ (g) $5 = \sqrt{4z + 1}$

(d) $2\sqrt{x} - 12 = 0$ (h) $3\sqrt{c - 6} = 9$

- - - - - - - - - - - - - -

(a) $x = 4$; check: $\sqrt{4} = 2$, $2 = 2$;
(b) $x = 5$; check: $\sqrt{9} = 3$, $3 = 3$;
(c) $x = 1$; check: $\sqrt{1} = 1$;
(d) $x = 36$; check: $2\sqrt{36} - 12 = 0$, $0 = 0$;
(e) $a = 7$; check: $2\sqrt{7 + 2} = 6$, $2 \cdot 3 = 6$, $6 = 6$;
(f) $x = 15$; check: $\sqrt{30 + 6} - 6 = 0$, $0 = 0$;
(g) $z = 6$; check: $5 = \sqrt{24 + 1}$, $5 = 5$;
(h) $c = 15$; check: $3\sqrt{15 - 6} = 9$, $3 \cdot 3 = 9$, $9 = 9$

UNIT SEVEN

Linear and Fractional Equations and Formulas

Review Item	Ref Page	Example
1. An *equation* is a shorthand way of saying that two algebraic expressions are equal.	127	$3 + 4a + b = 2c + 9$ $5x + 2 = 3x + 8$
2. Many equations contain at least one literal term (letter) whose value is unknown at the beginning of the problem. Finding the value of the unknown that satisfies the equation is called *solving the equation.* The value found is called the *root* of the equation.	127	Problem: $a + 3 = 7$ (*a* is the unknown literal term.) Solution: $a = 7 - 3$ $\qquad\qquad a = 4$ Root: $\qquad 4$
3. Equations are of three types: • An *identity* is true for all values of the variable or states a numerical equality. • A *false statement* is false for all values of the variable or an incorrect numerical result. • A *conditional* equation is true only for a limited number of values of the variable.	127	$x + 3 = \dfrac{2x + 6}{2}$ $x = x + 3$ $2a + 1 = 7$ is true only for the value $a = 3$

Review Item	Ref Page	Example
4. The axioms of equality allow you to perform any of the following operations on an equation without changing its value: • *Add* the same amount to both sides. • *Subtract* the same amount from both sides. • *Multiply* both sides by the same (nonzero) number. • *Divide* both sides by the same (nonzero) number.	128	$x - 2 = 3$ $x - 2 + 2 = 3 + 2$ $x = 5$ $x + 2 = 3$ $x + 2 - 2 = 3 - 2$ $x = 1$ $x \div 2 = 4$ $(x \div 2) \cdot 2 = 4 \cdot 2$ $x = 8$ $2x = 4$ $2x \div 2 = 4 \div 2$ $x = 2$
5. The value of a factor in any term of an equation can be changed to 1 by multiplying by the reciprocal of that factor. *Remember:* The reciprocal of a whole number is 1 divided by that number; the reciprocal of a fraction is the fraction inverted.	129	Given: $\quad 3y = 9$ $(\frac{1}{3}) 3y = (\frac{1}{3}) 9^3$ $y = 3$ Given: $\quad \dfrac{y}{3} = 3$ $^1 3 \cdot \dfrac{y}{3} = 3 \cdot 3$ $y = 9$
6. An *inverse operation* is one that has the effect of undoing another operation.	130	See review item 4 for examples of inverse operations.
7. Solving an equation for x means finding the value for $+x$, never the value for $-x$. If your solution is in terms of $-x$, multiply or divide both sides by -1 to find the solution for $+x$.	131	If $-x = 6$, then multiplying both sides by -1, we get $(-1)(-x) = (-1)6, \; x = -6.$ or dividing both sides by -1, we get $\dfrac{-x}{-1} = \dfrac{6}{-1}, \; x = -6.$

Review Item	Ref Page	Example
8. To solve an equation in which the unknown appears in both members: • Rewrite the equation so that all terms containing the unknown are in the left member and all other terms are in the right member. • Combine like terms. • Solve by making the coefficient of the variable 1.	131	Given: $4m - 7 = 2m + 3$ $4m - 2m = 3 + 7$ $2m = 10$ $m = 5$
9. To check your solution of an equation, substitute the value found for the unknown back into the original equation, which should then reduce to an identity.	132	From review item 8, $m = 5$. Substituting this value in the original equation, $4m - 7 = 2m + 3$ we get $4 \cdot 5 - 7 \overset{?}{=} 2 \cdot 5 + 3$ or $\quad 20 - 7 \overset{?}{=} 10 + 3$ $13 \overset{\checkmark}{=} 13$
10. To solve equations containing parentheses, first remove the parentheses. Remember to change the signs of all terms enclosed in parentheses preceded by a minus sign.	132	$(6x + 3) - (4x + 14) = 5$ $6x + 3 - 4x - 14 = 5$ $2x = 5 + 11$ $2x = 16$ $x = 8$
11. To solve fractional equations whose terms have the same denominator, clear the fractions by multiplying each term of the equation by the common denominator.	133	$\dfrac{3y}{4} + y = \dfrac{7}{4}$ $4(\dfrac{3y}{4} + y) = 4(\dfrac{7}{4})$ $\dfrac{(4)3y}{4} + (4)y = \dfrac{(4)7}{4}$ $3y + 4y = 7; \ y = 1$

Review Item	Ref Page	Example
12. To solve fractional equations whose terms have different denominators, clear the fractions by multiplying each term by the lowest common denominator (LCD).	134	$\dfrac{a+3}{a} + \dfrac{4}{3} = \dfrac{23}{3a}$ LCD $= 3a$ $\dfrac{(3a)(a+3)}{a} + \dfrac{(3a)4}{3} = \dfrac{3a \cdot 23}{3a}$ $3a + 9 + 4a = 23$ $7a = 14$ $a = 2$
13. To find the LCD for fractional equations having binomial denominators: • Write each denominator as the product of prime factors. • Write the product of all the different prime factors. • Use the largest exponent required for any given prime factor.	135	Given: $\dfrac{x}{x+1} + \dfrac{5}{8} = \dfrac{5}{2(x+1)} + \dfrac{3}{4}$ $x+1;\ 2^3;\ 2(x+1);\ 2^2$ $8(x+1)$ $2^3(x+1)$
14. To solve for a variable when its square has been found, take the square root of both sides of the equation.	136	$M = \dfrac{3b^2}{2}$; solve for b. $b^2 = \dfrac{2M}{3}$ and $b = \sqrt{\dfrac{2M}{3}}$
15. A *formula* is an equation used to express a rule or relationship in concise form. To solve a problem using a formula: • Write down the formula. • Use the axioms of equality to get the unknown term on the left side of the equation and all others on the right. • Substitute known values for the variables on the right and solve.	137	Formula: $A = bh$; $A = 90,\ h = 18,\ b = ?$ $A = bh$ $b = \dfrac{A}{h}$ $b = \dfrac{90}{18}$ $b = 5$

UNIT SEVEN REFERENCES

1. When we say that two algebraic expressions are equal, we mean that they have the same value. For example, $3n = 12$ is just a quick way of saying that three times a certain number equals 12. Similarly, the equation $3n + 4n = 7n$ tells us that three times a certain number added to four times the same number is equal to seven times the number.

We identify the part of the equation to the left of the equal sign as the *left member* and the part to the right (not surprisingly) as the *right member*. We will use these terms frequently to refer to parts of the equation.

Identify the parts of the equation $3x - 4y = 18 + 2z$.

(a) Left member _____

(b) Right member _____

- - - - - - - - - - - - -

(a) $3x - 4y$; (b) $18 + 2z$

2. In the equation $3c = 18$, the solution is 6, because 6 times 3 equals 18. So we say that the number 6 *satisfies* the equation; 6 is the *root* of the equation. The root of an equation, then, is any number which, when substituted for the unknown letter, makes the two sides of the equation identical—or, as we sometimes say, reduces the equation to an identity.

Find the roots of the following equations.

(a) $5x = 25$

(b) $4b = 12$

(c) $x - 3 = 7$

(d) $\dfrac{x}{2} = 3$

(e) $t + 3 = 8$

(f) $z - 3 = 4$

(g) $2b = 7$

(h) $x + 3 = 7$

(i) $\dfrac{y}{3} = 7$

(j) $ab = c$ (solve for a)

- - - - - - - - - - - - -

(a) $x = 5$; (b) $b = 3$; (c) $x = 10$; (d) $x = 6$; (e) $t = 5$; (f) $z = 7$;

(g) $b = \dfrac{7}{2}$ or $3\dfrac{1}{2}$; (h) $x = 4$; (i) $y = 21$; (j) $a = \dfrac{c}{b}$

3. See if you can identify each of the following as an identity, a false statement, or a conditional equation.

(a) $2x + 5 = 6$ _____

(b) $3x = 2x + 3$ _____

(c) $2(x + 3) = 2x + 6$ _____

(d) $3x = 3x + 7$ _____

(e) $4a - 6a = -2a$ _____

(f) $x - 3 = x$ _____

------ ---------

(a) conditional equation (true only for $x = \dfrac{1}{2}$);   (b) conditional equation (true only for $x = 3$);   (c) identity;   (d) false statement ($3x$ cannot be equal to itself plus something more);   (e) identity;   (f) false statement

4.   All the equations we have talked about thus far are *first degree* or *linear* equations, because the exponent of the variable is no higher than 1. A *conditional* linear equation always has just one root, that is, there is one and only one value for the variable that will make the equation a true statement. In the remainder of this unit, we will use the word *equation* to mean a conditional linear equation.

In order to find the solution (or root) of an equation, we often have to change it to an *equivalent equation* which we can solve by inspection. An equivalent equation has all of the solutions of the given equation, and no more.

In the example below, the derived equations (2) and (3) and the original equation (1) are equivalent because 5 is the only root of each:

$$
\begin{aligned}
(1) \quad 3x + 1 &= x + 11 \\
(2) \quad 2x + 1 &= 11 \\
(3) \quad\quad 2x &= 10 \\
x &= 5
\end{aligned}
$$

The axioms stated in review item 4 may be summarized by the following statement:

> *If the same number is added to or subtracted from both members of an equation, or if both sides of an equation are multiplied or divided by any constant (except zero), the derived equation is equivalent to the original one.*

They also lead to the following somewhat simplified rule for transposing terms across the equal sign by means of addition or subtraction:

> *A positive term may be eliminated from one member of an equation by subtracting that term from each member of the equation. A negative term may be eliminated from one member of an equation by adding that term to both members of the equation.*

Apply this rule in solving the following equation:

$$7 + y = 16$$
$$7 - \underline{\quad} + y = 16 - \underline{\quad}$$
$$y = \underline{\quad}$$

- - - - - - - - - - - - - -

$7 - 7 + y = 16 - 7;\ y = 9$

5. Variables in equations are sometimes connected with other numbers that are factors or divisors of that variable. Consider, for example, the relationship among the time (T) a moving object travels, the speed at which it travels (S), and the distance it travels (D). Distance is equal to time multiplied by speed, which gives us the relationship $D = T \cdot S$, or $D = TS$.

Now, suppose we have a problem in which we know the values for distance and speed but wish to determine the value of time. We must rewrite the equation so that T (the unknown factor) appears by itself on one side of the equal sign, with the two known values, S and D, on the other side. Dividing both sides by S we get:

$$\frac{D}{S} = \frac{T \cdot \cancel{S}}{\cancel{S}} \qquad \left(\textit{Note: } \frac{S}{S} = 1\right)$$

Interchanging members to get T on the left side, we get:

$$T = \frac{D}{S}$$

Had we started with $T = \dfrac{D}{S}$ and wanted to solve for D, we would have multiplied both sides of the equation by S, thus canceling the S in the denominator of the right member and making it a multiplier of T in the left member:

$$T \cdot S = \frac{D \cdot \cancel{S}}{\cancel{S}} \qquad \text{or} \qquad D = TS \qquad \text{the second equation being our}$$
original equation.

The two convenient rules that follow are simply a somewhat more formal statement of the rule given in review item 5:

- *A factor in any member of an equation may be changed to a value of 1 by dividing both members of the equation by that factor.*
- *A divisor of any term in an equation can be changed to a value of 1 by multiplying each member of the equation by that divisor.*

Apply these rules in solving the equation $\dfrac{6}{a} = 3$.

- - - - - - - - - - - - -

Multiplying both sides by a gives us:

$$\frac{6\cancel{a}}{\cancel{a}} = 3a$$

$$3a = 6$$

Dividing both sides by 3 gives us:

$$\frac{\cancel{3}a}{\cancel{3}} = \frac{\cancel{3} \cdot 2}{\cancel{3}}$$

From which we conclude:

$$a = 2$$

6. *Inverse operations* are the kind of operations we have been using in the examples for review items 4 and 5. Only the name is new. Addition and subtraction are inverse operations. So are multiplication and division. For example, we simplify the equation $2k + 3 = 7$ by subtracting 3 from both members. We subtract 3 because this is the inverse of adding 3, and we want to eliminate the 3 that was added to the $2k$. Doing this yields the equivalent equation $2k = 4$, which we solve by dividing both members by 2. We divide by 2 because this is the inverse of multiplication by 2, and we want to change the factor of 2 to a factor of 1. Before you try some problems on your own, remember:

- *Perform addition to undo subtraction, and subtraction to undo addition, when you wish to eliminate a number (that is, make it equal to zero).*

- *Perform multiplication to undo division, and division to undo multiplication, when you wish to make a multiplier or divisor become 1.*

In each of the following problems, you must decide whether to add a positive number or a negative number to both sides. To solve these equations, make a term become zero by adding its opposite.

		Term to be added to each side	Answer
(a)	$x - 5 = 10$	_____	$x = $ ___
(b)	$b - 7 = 7$	_____	$b = $ ___
(c)	$2 = 1 - m$	_____	$m = $ ___
(d)	$6 = 6 - y$	_____	$y = $ ___
(e)	$2y + 3 = 9$	_____	$y = $ ___

Use the rule that both sides of an equation may be multiplied by the same non-zero quantity in solving the following equations.

(f) $\dfrac{y}{2} = 6$

(g) $5 = \dfrac{k}{5}$

(h) $\dfrac{z}{9} = -2$

(i) $\dfrac{a}{b} = c \ (b \neq 0)$

(j) $\dfrac{1}{3} = \dfrac{x}{2}$

------ ---------

(a) 5, $x = 15$;   (b) 7, $b = 14$;   (c) –2, $m = -1$ (see reference item 7 below);
(d) –6, $y = 0$;   (e) –3, $y = 3$;   (f) $y = 12$;   (g) $k = 25$;   (h) $z = -18$;
(i) $a = bc$;   (j) $x = \dfrac{2}{3}$

7. To convert the solution of an equation in terms of $-x$ into $+x$, we multiply, or divide, both sides of the equation by –1. If your solution turned out to be $-x = 8$, multiplying or dividing both sides by –1 would give $x = -8$. If the solution was $-x = -8$, multiplication or division by –1 would give $x = 8$.

Below is a mixed set of problems that will give you opportunities to apply the rules we have been reviewing. Solve the following equations.

(a) $x + 7 = 10$

(b) $3x - x = 6$

(c) $2 = k - 4$

(d) $2p - 6 = 4$

(e) $3 - 9 = 5r - 2r$

(f) $2y - 3y - 1 = 2$

(g) $-3 = \dfrac{y}{2}$

(h) $6m - 11 = 1$

(i) $\dfrac{-3z}{4} = -6$

(j) $4 - \dfrac{2x}{5} = 6$

------ ---------

(a) $x = 3$; (b) $x = 3$; (c) $k = 6$; (d) $p = 5$; (e) $r = -2$; (f) $y = -3$; (g) $y = -6$;
(h) $m = 2$; (i) $z = 8$; (j) $x = -5$

8. In the examples and problems we have used so far, the terms containing the letters have all been in one member—usually the left member. Review item 8 describes the procedure for solving equations in which the unknown appears in both members. Here is another example. Suppose we have the equation $6x - 2 = 8 + 4x$. Adding $+2$ and $-4x$ to each side gives us $6x - 4x = 8 + 2$. Combining like terms, we get $2x = 10$, and multiplying both sides by $\frac{1}{2}$ (to make the coefficient of the variable 1) gives us $x = 5$. To check our results, we substitute 5 for x in the original equation, giving us $6(5) - 2 \overset{?}{=} 8 + 4(5)$, or $28 \overset{\checkmark}{=} 28$.

[Note: The equal sign with the question mark over it ($\overset{?}{=}$) and the equal sign with the check mark over it ($\overset{\checkmark}{=}$) are often used to indicate the checking process.]

Solve and check the following equations.

(a) $3a - 5 = 16$

(b) $3x = 7 - 4x$

(c) $6x + 21 = 84 - 3x$

(d) $7y + 28 = 5y + 6$

(e) $16 = 7 - 3x$

(f) $8x - 9 - 5x = 0$ (Do not be afraid of the zero; just add 9 to both sides.)

(g) $5r + 12 - 3r = -2 - 13 - r$

(h) $11c - 8 - 2c - 64 = 0$

- - - - - - - - - - - - - -

(a) $a = 7$, $16 \overset{\checkmark}{=} 16$; (b) $x = 1$, $3 \overset{\checkmark}{=} 3$; (c) $x = 7$, $63 \overset{\checkmark}{=} 63$;

(d) $y = -11$, $-49 \overset{\checkmark}{=} -49$; (e) $x = -3$, $16 \overset{\checkmark}{=} 16$; (f) $x = 3$, $0 \overset{\checkmark}{=} 0$;

(g) $r = -9$, $-6 \overset{\checkmark}{=} -6$; (h) $c = 8$, $0 \overset{\checkmark}{=} 0$

9. The problems in reference item 8 gave you ample opportunity to practice the checking procedure shown in review item 9. The important thing is that you get in the habit of *always* checking your solution to an equation, and *always* by substituting the root you have found by inserting it in the original equation, making sure it reduces to an identity.

10. In Unit 4, you were given some practice in removing parentheses and other grouping symbols that will prove helpful to you now. Consider, for example, the equation $(7x + 5) - (2x - 15) = 10$. The first step in solving this equation is to remove the parentheses, changing signs of all terms in the subtrahend, since it is preceded by a minus sign:

$$7x + 5 - 2x + 15 = 10$$

Now we can complete the solution:

$$
\begin{aligned}
7x - 2x &= 10 - 5 - 15 \\
5x &= -10 \\
x &= -2
\end{aligned}
$$

We check the solution by substituting –2 for x in the original equation:

$$
\begin{aligned}
(-14 + 5) - (-4 - 15) &\overset{?}{=} 10 \\
-9 - (-19) &\overset{?}{=} 10 \\
-9 + 19 &\overset{?}{=} 10 \\
10 &\overset{\checkmark}{=} 10
\end{aligned}
$$

Use this example as guidance to help you solve and check the following.

(a) $5(b + 4) - 4(b + 3) = 0$

(b) $3x - 4(x + 2) = 5$

(c) $-(5x + 4) = -6x + 3$

(d) $4(a - 3) - 6(a + 1) = 0$

(e) $5(2x + 3) = 2(4x - 1)$

(f) $7(p - 5) = 14 - (p + 1)$

(g) $(m - 9) - (m + 7) = 4m$

(h) $20 = 8 - 2(9 - 3x)$

Solve.

(i) $7(5t - 1) - 18t = 12t - (3 - t)$

(j) $5x - 4(x - 6) - 2(x + 6) + 12 = 0$

(k) $\frac{1}{3}(6y - 9) = \frac{1}{2}(8y - 4)$

(l) $24b^2 + 147 = 27 - 16b - 4b(1 - 6b)$

- - - - - - - - - - - - - -

(a) $b = -8, 0 \overset{\checkmark}{=} 0$; (b) $x = -13, 5 \overset{\checkmark}{=} 5$; (c) $x = 7, -39 \overset{\checkmark}{=} -39$;

(d) $a = -9, 0 \overset{\checkmark}{=} 0$; (e) $x = -\frac{17}{2}, -70$; (f) $p = 6, 7 \overset{\checkmark}{=} 7$;

(g) $m = -4, -16 \overset{\checkmark}{=} -16$; (h) $x = 5, 20 \overset{\checkmark}{=} 20$; (i) $t = 1$; (j) $x = 24$;

(k) $y = -\frac{1}{2}$; (l) $b = -6$

11. Equations that contain fractions can usually be solved more easily if they are changed into equations that do not contain fractions. To do this, multiply each member of the equation by the smallest number that is exactly divisible by each denominator—the lowest common denominator (LCD). This process is known as *clearing fractions*.

We reviewed this idea briefly in review item 6 in order to solve equations of the type $\frac{x}{a} = c$. Now we will consider equations with more than one literal term, one or more of which contains a fraction with the same denominator.

Example 1: $\frac{3y}{7} - 2 = \frac{y}{7}$

Solution: Multiply all terms by 7:

$$\frac{(7)3y}{7} - (7)2 = \frac{(7)y}{7}$$

$$3y - 14 = y$$
$$y = 7$$

Example 2:

$$\frac{4}{x} = 5 - \frac{1}{x}$$

Solution: Multiply all terms by x:

$$\frac{(x)4}{x} = (x)5 - \frac{(x)1}{x} \qquad (x \neq 0)$$

$$4 = 5x - 1$$
$$5x = 5$$
$$x = 1$$

Use the rule given in review item 11 and illustrated in the examples above to solve the following equations by clearing fractions.

(a) $\dfrac{2x}{3} + 5 = 5$

(b) $\dfrac{x}{3} - x = 2$

(c) $10 - \dfrac{3y}{5} = y - 6$

(d) $\dfrac{3}{r} = 2 - \dfrac{7}{r}$

(e) $\dfrac{10}{x} - 2 = 18$

(f) $10 - \dfrac{z}{7} = \dfrac{4z}{7}$

(g) $\dfrac{6h}{5} + 8 = \dfrac{2h}{5}$

(h) $\dfrac{3y - 1}{7} = 2y + 3$

(a) $x = 3$; (b) $x = -3$; (c) $y = 10$; (d) $r = 5$; (e) $x = \dfrac{1}{2}$; (f) $z = 14$;

(g) $h = -10$; (h) $y = -2$

12. Here are two more examples of the procedure for solving equations having different denominators. The first example has simple numerical denominators in the two fractions; the second contains the unknown in the denominators of the two fractions.

Example 1: $\dfrac{x}{2} + \dfrac{x}{3} = 5$

Solution: LCD $= 2 \cdot 3 = 6$

Multiply all terms by 6:

$$\frac{(6)x}{2} + \frac{(6)x}{3} = (6)5$$

$$3x + 2x = 30$$

$$x = 6$$

Example 2:

$$\frac{10}{x} = \frac{25}{3x} - \frac{1}{3} \qquad \text{*Solution:* LCD} = 3x$$

Multiply all terms by $3x$:

$$\frac{(3x)10}{x} = \frac{(3x)25}{3x} - \frac{(3x)1}{3}$$

$$3 \cdot 10 = 25 - x$$

$$x = -5$$

Solve the following equations.

(a) $\dfrac{3y}{4} - \dfrac{2y}{3} = \dfrac{3}{4}$ LCD = _____ y = _____

(b) $\dfrac{x}{4} + \dfrac{x}{3} + \dfrac{x}{2} = 26$ LCD = _____ x = _____

(c) $\dfrac{5}{x} - \dfrac{2}{x} = 3$ LCD = _____ x = _____

(d) $\dfrac{3}{4c} = \dfrac{1}{c} - \dfrac{1}{4}$ LCD = _____ c = _____

(e) $\dfrac{b-6}{b} = \dfrac{10}{7}$ LCD = _____ b = _____

— — — — — — — — — — — — —

(a) LCD = 12, $y = 9$; (b) LCD = 12, $x = 24$; (c) LCD = x, $x = 1$;

(d) LCD = $4c$, $c = 1$; (e) LCD = $7b$, $b = -14$

13. Finding the LCD for fractional equations that have binomial denominators is a bit tricky, although the procedure is essentially the same as the one we used in review items 11 and 12. Here is another example to help familiarize you with the technique:

Example: Find the LCD for $\dfrac{2}{x^2 - 4}$, $\dfrac{3}{4x - 8}$, and $\dfrac{4}{x^2 + 4x + 4}$.

Solution: Write each denominator as the product of prime factors:

$$x^2 - 4 = (x + 2)(x - 2)$$

$$4x - 8 = 4(x - 2)$$

$$x^2 + 4x + 4 = (x + 2)^2$$

Write the product of all the different prime factors:

$4(x - 2)(x + 2)$

Use the largest exponent required with each factor in the product. The result is the LCD:

$4(x - 2)(x + 2)^2$

It is wise always to test to make sure that the LCD is divisible by each of the denominators.

Now for a little practice. Find the LCD of each of the following sets of fractions.

(a) $\dfrac{3}{2x - 6}$ $\dfrac{4}{x^2 - 9}$ $\dfrac{18}{6x + 18}$ LCD = _____

(b) $\dfrac{6}{x^2 - 1}$ $\dfrac{7}{x^2 - 2x + 1}$ $\dfrac{9}{x + 1}$ LCD = _____

(c) $\dfrac{7}{x + 4}$ $\dfrac{2}{x - 3}$ $\dfrac{5}{x - 2}$ LCD = _____

Solve the following equations by first multiplying each term by the LCD of the fractions in the equation.

(d) $\dfrac{a-2}{3} - \dfrac{a + 1}{4} = 1$

(e) $\dfrac{x + 8}{x - 2} = \dfrac{9}{4}$

(f) $\dfrac{4}{3(5 + k)} - \dfrac{2}{3} = \dfrac{4}{5 + k}$

(g) $\dfrac{8}{x - 2} - \dfrac{13}{2} = \dfrac{3}{2x - 4}$ *

(h) $\dfrac{3}{a + 2} + \dfrac{a - 2}{4} = \dfrac{a - 3}{4}$

(i) $\dfrac{4}{x - 4} = \dfrac{7}{x + 2}$

- - - - - - - - - - - - - -

(a) $6(x^2 - 9)$ or $2 \cdot 3(x - 3)(x + 3)$; (b) $(x - 1)^2 (x + 1)$; (c) $(x + 4)(x - 3)$ $(x - 2)$; (d) $a = 23$; (e) $x = 10$; (f) $k = -9$; (g) $x = 3$; (h) $a = -14$;

(i) $x = 12$

14. Occasionally you will have to work with equations in which the letter you wish to solve for is squared. For example, if we wished to solve the equation

---

*Write this in factored form before trying to find the LCD.

$F = \dfrac{mv^2}{2}$ for v, we would first multiply both sides by $\dfrac{2}{m}$, giving us $\dfrac{2F}{m} = v^2$ or,

interchanging the right and left members, $v^2 = \dfrac{2F}{m}$. Since we wish to solve for

v, not v^2, the final step in solving for v would be to take the square root of both sides, giving us

$$v = \sqrt{\dfrac{2F}{m}}$$

This tells us that to find the value of v, we first have to multiply the value of F by 2, divide the result by the value of m, and take the square root of the resulting number. Thus, if $F = 54$ and $m = 3$, $v = \sqrt{\dfrac{2 \cdot 54}{3}} = \sqrt{36} = 6$.

Practice transforming and evaluating equations by solving the following.

(a) Find b if $A = bh$; $A = 48$, $h = 8$.

(b) Find d if $Q = 2d + k$; $Q = 9$, $k = 3$.

(c) Find E if $I = \dfrac{E}{R + r}$; $I = 15$, $R = 9$, $r = 3$.

(d) Find E if $C = \dfrac{nE}{R + nr}$; $C = 3$, $R = 6$, $n = 2$, $r = 4$.

(e) Find e if $T = 6e^2$; $T = 96$.

- - - - - - - - - - - - - - -

(a) $b = \dfrac{A}{h}$, $b = 6$; (b) $d = \dfrac{Q - k}{2}$, $d = 3$; (c) $E = I(R + r)$, $E = 180$;

(d) $E = \dfrac{C(R + nr)}{n}$, $E = 21$; (e) $e = \sqrt{\dfrac{T}{6}}$, $e = 4$

15. Mathematical formulas are used to find the circumference of a circle, the height of a triangle, the volume of a cube, the speed of a moving vehicle, the gravitational pull between the earth and the moon, the amount of energy in a unit of mass, and so on. The uses of formulas are literally endless. We worked with one such formula in review item 5: the time-speed-distance relationship, expressed as $D = TS$ (that is, distance is equal to speed multiplied by time, or vice versa.) Using our inverse operations rules, we were able to express any one of the three quantities (T, D, or S) in terms of the other two. Thus, by using division to undo multiplication, we could divide both numbers by S to get $T = D/S$, or by T to get $S = D/T$.

In the example for review item 15 we use the formula $A = bh$, where A is the variable we are solving for—also known as the *subject*. This particular formula says that the area A of a rectangular surface is equal to the length of the base b multiplied by the height h.

Suppose, now, that we wished to change the subject to h, that is, to express the height in terms of the other two quantities. Since the original form of the equation is $A = bh$, we simply have to divide both sides by b, or multiply both sides by $\frac{1}{b}$, which gives us $h = \frac{A}{b}$.

Use the axioms from review item 4 to transform the following formulas as required.

(a) Find k if $Z = \dfrac{m^2 k}{4}$; $k = $ _____

(b) Find I if $W = \dfrac{1}{2} LI^2$; $I = $ _____

(c) Find P if $X = \dfrac{M}{N} P$; $P = $ _____

(d) Find V if $R = \dfrac{pVb}{M}$; $V = $ _____

- - - - - - - - - - - - - -

(a) $k = \dfrac{4Z}{m^2}$;   (b) $I = \sqrt{\dfrac{2W}{L}}$ (see reference item 14);

(c) $P = \dfrac{N}{M} X$;   (d) $V = \dfrac{RM}{pb}$

UNIT EIGHT

Functions and Graphs

Review Item	Ref Page	Example
1. A *rectangular coordinate system* is used to represent equations in graph form. As shown at the right, it is formed by combining two number scales at right angles to each other in such a way that their zero points coincide at the *point of origin.* The horizontal scale is called the *X axis;* the vertical scale is called the *Y axis.*	143	
2. The *X* axis and *Y* axis divide the graph area into four parts, called *quadrants.* These are numbered from I to IV starting with the upper right quadrant and proceeding counter‑clockwise. Points are located and identified according to their distance from the *X* and *Y* axes. These distances determine the *coordinates* of the point. The *x* coordinate is called the *abscissa;* the *y* coordinate is called the *ordinate.*	144	

139

Review Item	Ref Page	Example
3. The sets of coordinates that serve to locate points on a plane (i.e., flat) surface are called *ordered pairs* because they always appear in the same order: x value first, y value second.	144	In the example for review item 2 above, (−2, −4) and (3, 2) are ordered pairs.
Ordered pairs derive primarily from equations or inequalities which describe a particular relationship between two unknowns (usually x and y) in such a way that the value of one depends upon the value of the other.		In the equation $y = 3x − 2$, if $x = 0$, then $y = −2$; if $x = 1$, then $y = 1$; if $x = 2$, then $y = 4$; and so on.
4. In a relationship such as $y = 2x$, because x and y can assume different values depending upon the values assigned to the other, x and y are called *variables*. We can also refer to the relation $y = 2x$ as a *function*, and say that y is a function of x.	145	In the equation $y = 2x$, x and y are the variables.
The variable that has numerical values assigned to it is referred to as the *independent variable*.		In the equation above, x is the independent variable.
The variable whose value is determined by the values assigned to the other variable is referred to as the *dependent variable*.		y is the dependent variable because its value depends on the values assigned to x.
5. In a *first degree* or *linear equation*, the exponents of the variables are no higher than 1. To graph, or *plot*, a linear equation:	145	See reference item 5.

Review Item	Ref Page	Example
• Rearrange the equation so that one unknown is alone on the left and all other terms are on the right. • Assign at least three values to the unknown on the right, and find the corresponding values of the unknown on the left. • Plot the resulting pairs of coordinates, and connect the points with a straight line.		
6. The coordinates of any point on the graph of an equation *satisfy* the equation, that is, they make it a true statement.	147	From reference item 5, substituting the coordinate pair (–1, –2) in the equation $y \overset{?}{=} 2x$, we get $-2 = 2(-1)$ or $-2 \overset{\checkmark}{=} -2$, a true statement.
7. *Intercepts* are the points at which the graph of an equation intersects (crosses) the x and y axes, that is, the points at which y and x, in turn, equal zero.	147	In the equation $2x + y - 4 = 0$, when $x = 0$, $y = 4$; and when $y = 0$, $x = 2$. Thus, $y = 4$ and $x = 2$ are the intercepts.
8. The graph of a first degree equation in one unknown is either the X axis, the Y axis, or a line parallel to one of these.	148	See reference item 8.
9. To solve linear equations for unique values of the unknowns, you need as many equations as there are unknown quantities.	149	Unique values of x and y cannot be found from the single equation $2x + y = 7$. But with the additional equation $3x - y = 8$, we can find unique values of x and y.

Review Item	Ref Page	Example
10. Linear equations can be solved graphically by plotting both equations on the same coordinate system and finding the point at which the two lines intersect. The coordinates of this point represent the x and y values that satisfy both equations.	149	See reference item 10.
11. Equations that intersect one another have one pair of values that satisfies both of them. They are called *consistent* equations. Equations are termed *inconsistent* if no pair of values satisfies both of them.	151	See reference item 11.
12. *Dependent equations* are two equations which plot as the same straight line. The terms of one dependent equation are simply multiples of the other.	151	$2x + y = 3$ and $4x + 2y = 6$ are dependent equations.
13. Linear equations can be solved by adding or subtracting.	152	Adding: $$\begin{array}{r} 2x + y = 7 \\ 3x - y = 8 \\ \hline 5x \quad\;\; = 15 \\ x = 3 \end{array}$$ using the value for x in one of the original equations: $2 \cdot 3 + y = 7$ $y = 1$
14. Linear equations can also be solved by substitution, that is, rewriting one equation in terms of one of the unknowns, and substituting this value in either of the two equations	153	Given: (1) $2x + y = 7$ (2) $3x - y = 8$ From (1): $y = 7 - 2x$. Substituting this value of y in equation (2), we get: $3x - (7 - 2x) = 8$ from which $x = 3$; $y = 1$.

UNIT EIGHT REFERENCES

1. You will recall that an equation is a statement that two expressions are equal. In this unit, we will be working with a particular type of equation—the *first degree,* or *linear,* equation in one or two unknowns (variables). This means that there will be, at most, two variables in any given equation, and no variable will have an exponent greater than 1.

In order to plot points in two dimensions (i.e., on a plane surface), also known as *graphing an equation,* we need a coordinate system. The coordinate system we use is known as a *rectangular coordinate system* because, as indicated in review item 1, it is formed by combining two number scales at right angles to each other. Their zero points coincide at a point known as the *point of origin,* or simply the *origin.* Observe this in the graph below.

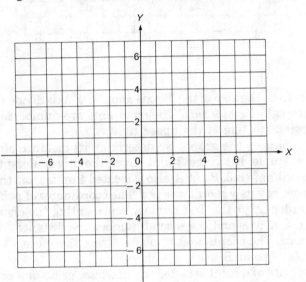

The horizontal scale is called the *X axis*; the vertical scale is called the *Y axis.* The origin, 0, is the point where the two scales intersect.

(a) The vertical scale is called the _____ .

(b) The horizontal scale is called the _____ .

(c) The point where the two scales intersect is known as the

_____

(a) *Y* axis;   (b) *X* axis;   (c) origin

2.

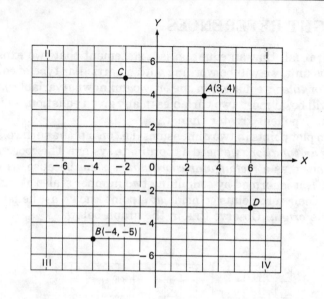

As you can see from the above, the X axis and the Y axis divide the graph area into four parts, called *quadrants*. The quadrants are numbered from I to IV, counterclockwise, starting at the upper right.

To locate a point on the graph, we determine its distance along each axis. Point A, for example, is located 3 units from the origin along the X axis; so we say its x coordinate is 3. Point A also is located 4 units from the origin along the Y axis; so we say its y coordinate is 4. The coordinates of points are always listed in the order (x, y), the first number representing the distance from the origin along the X axis and the second number the distance from the origin along the Y axis. The coordinates of point A are, therefore, (3, 4). Similarly, the coordinates of point B are (– 4, –5).

The x coordinate of a point is called the *abscissa,* or its *first component.* The y coordinate of a point is called the *ordinate,* or its *second component.*

In the graph just shown, what are the coordinates of points C and D?

(a)   The coordinates of C are _____ .

(b)   The coordinates of D are _____ .

(a) (–2, 5);   (b) (6, –3)

3.   What mathematicians call a *relation* between two variables, usually defined by an equation such as $y = 2x$, results in a set of *ordered pairs.* If we assign a whole series of possible values to x, the relation $y = 2x$ will give us a whole series of corresponding values for y. Thus, we have a set of ordered pairs. Both equations and inequalities, which we will get into later, can be used to form ordered pairs.

One way (often the best) of solving equations and inequalities is by graphing them on a rectangular coordinate system. And graphing requires the use of ordered pairs.

Write the special names for each of the coordinates in an ordered pair.

(a) The first (x) coordinate is called the _____ .

(b) The second (y) coordinate is called the _____ .

- - - - - - - - - - - - - -

(a) abscissa (or first component);   (b) ordinate (or second component).

4.   From review item 4, you know that x and y in a relation such as $y = 2x$ are called *variables*. You also know that we can refer to the relation $y = 2x$ as a *function*, and say that y is a function of x because, as we assign various values to x, y takes on different, corresponding values.

From reference item 15, Unit 7, you know that whichever variable we solve an equation for—usually the left member of the equation—is called the *subject.* In an equation such as $y = 2x - 3$, y is the subject. It also is the dependent variable, since its value(s) will depend upon the values assigned to x.

To make sure you are clear about these relationships, answer the following questions regarding the equation $y = 3x + 5$. Bear in mind that the subject of the equation is the second component.

(a) In order to plot this equation, you would be forming ordered pairs in the form _____ .

(b) ____ would be the independent variable.

(c) ____ would be the dependent variable.

(d) ____ is a function of ____.

- - - - - - - - - - - - - -

(a) (x, y);   (b) x (x is the first component; therefore, it would be more natural to assign values to it);   (c) y (y is the second component; therefore, the values of y are determined by, or depend on, the values of x);   (d) y, x (the second component is a function of the first)

5.   Now you are ready to try graphing an equation. Let us use the relation $y = 2x$. First, you need to assign at least three values to x to see what corresponding values result for y. In this way, you will develop three or more ordered pairs which, when plotted, will represent the graph of the equation $y = 2x$.

The partially completed table below contains a series of x values. Substitute each of these values in the given equation ($y = 2x$), and write down the corresponding y values in the spaces provided. Finally, plot each pair of coordinates, and connect the resulting points with a straight line in order to find out what the graph of $y = 2x$ looks like. (We know in advance that it will have to be a straight line, because first degree equations are, by definition, linear, that is, they produce a straight line when plotted.)

x	3	2	1	0	−1	−3
y						

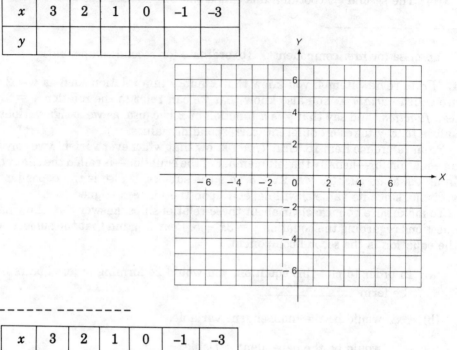

x	3	2	1	0	−1	−3
y	6	4	2	0	−2	−6

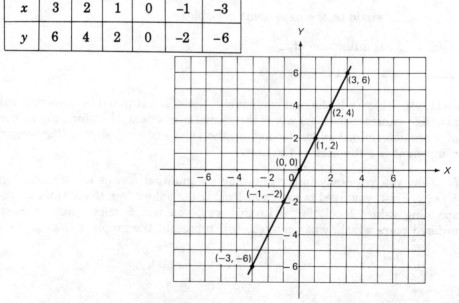

6. Here is another example to illustrate the fact that the coordinates of any point on the graph of an equation satisfy the equation, that is, make the equation a true statement.

Consider the equation $y = x + 2$, from review item 7. If we substitute the coordinate pair (2, 4) in this equation, we get

$$y \overset{?}{=} x + 2$$
$$4 = 2 + 2$$
$$4 \overset{\checkmark}{=} 4 \quad \text{(a true statement)}$$

If we substitute the coordinate pair (–2, 0), we get

$$0 = -2 + 2$$
$$0 = 0 \quad \text{(a true statement)}$$

Suppose, however, we substitute the coordinates of a point not on the graph. What do you think the result will be? Try it and see. Substitute the coordinate pair (3, – 4) in the equation $y = x + 2$ and see what you get.

------ --------

$y = x + 2; \quad -4 = 3 + 2; \quad -4 = 5$ (not a true statement)

7. Below is a graph of the equation $y = x + 2$.

x	2	0	-2
y	4	2	0

In this example, notice that when x is zero, y is 2, and when y is zero, x is –2. If the x value of a point is equal to zero, the point lies on the Y axis; if the *y* value of a point is equal to zero, the point lies on the *X* axis. This means that the graph of the equation intersects, or crosses, the axes at these points.

These special points, which we call *intercepts,* can provide a quick way of locating at least two points on the graph. The procedure for finding the *x* and *y* intercepts is as follows:

> To find the *x* intercept, set *y* equal to zero and solve the resulting equation for *x*. To find the *y* intercept, set *x* equal to zero and solve the resulting equation for *y*.

Example: Find the *x* and *y* intercepts of the equation $4x - 2y = 8$.

Solution: Setting $y = 0$: $4x - 0 = 8$
$$x = 2; \; x \text{ intercept } (2, 0)$$

Setting $x = 0$: $0 - 2y = 8$
$$y = -4; \; y \text{ intercept } (0, -4)$$

Apply this procedure to the following problems.

	x intercept	*y* intercept
(a) $3x + y = 9$	(___, 0)	(0, ___)
(b) $y - 5x = 10$	(___, 0)	(0, ___)
(c) $x + y = -3$	(___, 0)	(0, ___)
(d) $2y - 4x + 8 = 0$	(___, 0)	(0, ___)

- - - - - - - - - - - -

(a) (3, 0), (0, 9); (b) (–2, 0), (0, 10);   (c) (– 3, 0), (0, – 3);   (d) (2, 0), (0, – 4)

8. We have been considering the graphing of linear equations in two unknowns. What would the graph of a linear equation of one unknown look like?

Consider the equation $y = 2$. Can you picture the graph of this equation on the rectangular coordinate system? It would be a straight line composed of a series of points all 2 units above the *X* axis and, therefore, *parallel to* the *X* axis.

How about the graph of $x = 0$? It would simply *be* the *Y* axis.

Plot the following equations and test the validity of the rule given in review item 8.

(a) $x = -4$ (d) $y = 3\frac{1}{2}$

(b) $y = 0$ (e) $x = 0$

(c) $x = 2$ (f) $y = -5$

------- --------

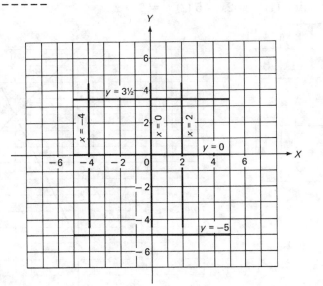

9. In Unit 7 we reviewed the procedure for solving linear equations in one unknown. How do you find a single solution to an equation containing two unknowns? The answer is that *it cannot be done.* As review item 9 indicates, to solve linear equations for unique values of the two unknowns, you must have as many equations (that is, relationships between the unknowns) as there are unknown quantities.

Thus, if we have three unknowns, we will need three different equations containing these unknowns to find a unique solution for each of the three unknowns. Five unknowns would require five different equations, and so on. For the remainder of this unit, we are going to discuss the solution of pairs of linear equations in two unknowns. In review item 10, we will talk about graphic solutions, and in review items 13 and 14 we will consider algebraic solutions.

10. *Example:* Solve the following pair of linear equations graphically.

$$2x - y = 5$$
$$x + y = 7$$

Solution: Since these two equations will have a common solution, that is, a pair of values for x and y that will satisfy both equations, the point represented by that pair of values must lie on the line graphs of both equations. The common solution must, therefore, be represented by the point at which the two lines intersect.

Let us see if this is so. To plot the two equations, we will arrange each in terms of y, develop at least three pairs of coordinates for each, and plot the resulting pairs: (1) $y = 2x - 5$ (2) $y = 7 - x$

x	0	2	4
y	−5	−1	3

x	0	2	4
y	7	5	3

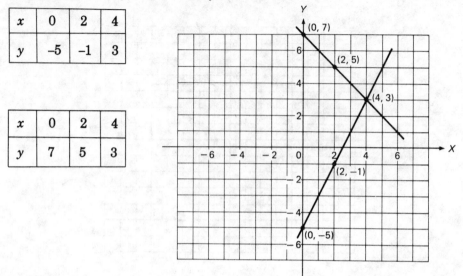

Since the two lines intersect at the point (4, 3), $x = 4$ and $y = 3$ is the common solution.

Using a sheet of graph paper, follow the same procedure to solve this equation pair: (1) $2y - x = 2$;   (2) $y + 2x = 6$.

----

(1) $2y - x = 2$, $y = \dfrac{2 + x}{2}$;   (2) $y + 2x = 6$, $y = 6 - 2x$; $x = 2$, $y = 2$ is the common solution.

x	−2	0	4
y	0	1	3

x	0	2	4
y	6	2	−2

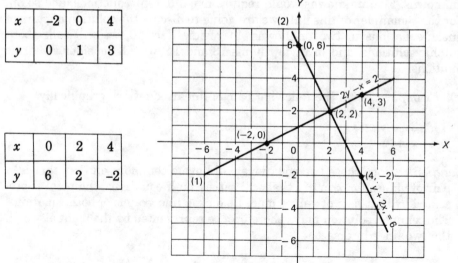

11. The pair of equations in reference item 10 would be termed *consistent equations,* since they have a common point of intersection, that is, one pair of values that satisfies them both. To see what a pair of *inconsistent equations* looks like, graph the following pair of equations: (1) $x + y = 4$; (2) $x + y = 6$.

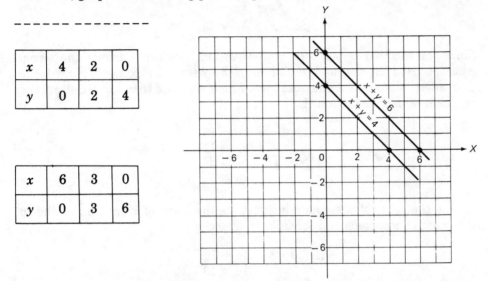

x	4	2	0
y	0	2	4

x	6	3	0
y	0	3	6

As you can see from the graph, inconsistent equations are simply parallel straight lines that cannot meet (within the system of Euclidean geometry, at least); hence, they have no common solution. Such equations are inconsistent for the obvious reason that two given numbers (represented by x and y) cannot add up to two different answers (in this case, 4 and 6).

12. Because *dependent equations* are simply multiples of each other, they are equivalent and produce the same graph. Prove this to yourself by graphing these two equations: (1) $y = x + 3$ and (2) $2y = 2x + 6$.

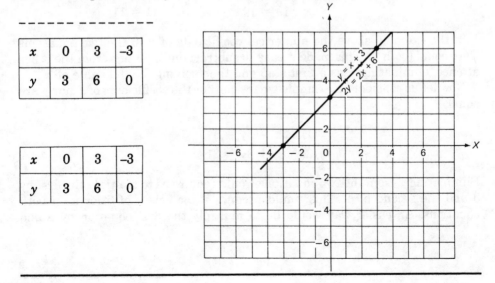

x	0	3	-3
y	3	6	0

x	0	3	-3
y	3	6	0

13. Now it is time to review the algebraic methods for simultaneously solving systems (i.e., pairs) of linear equations in two variables. Consider, for example, this system of equations:

$$(1) \quad 5x + 3y = 19$$
$$(2) \quad x + 3y = 11$$

Since the term $3y$ appears in both equations, we can subtract the second equation from the first and so eliminate the y term altogether. This will produce an equation in just one unknown, which we could then solve. Subtracting equation (2) from equation (1):

$$(1) \quad 5x + 3y = 19$$
$$(2) \quad \underline{ x + 3y = 11}$$
$$4x = 8$$
$$x = 2$$

To find the value of y, we merely substitue the value of x back into either of the original equations and solve for y. Thus:

$$(1) \quad 5 \cdot 2 + 3y = 19$$
$$3y = 19 - 10$$
$$3y = 9$$
$$y = 3$$

To check our solution, we substitute the values for x and y back into each of the original equations:

$$(1) \; 5 \cdot 2 + 3 \cdot 3 \overset{?}{=} 19 \qquad (2) \; 2 + 3 \cdot 3 \overset{?}{=} 11$$
$$10 + 9 \overset{?}{=} 19 \qquad\qquad 2 + 9 \overset{?}{=} 11$$
$$19 \overset{\checkmark}{=} 19 \qquad\qquad 11 \overset{\checkmark}{=} 11$$

In the example above the numerical coefficients of y in the two equations were equal (both 3). This made it easy to subtract one equation from the other at once, eliminating y as a first step and then solving for the value of x.

Now let us consider an example where neither the coefficients of x nor y are equal.

$$(1) \quad 2x - y = 5$$
$$(2) \quad 3x + 3y = 21$$

Eliminating x from the two equations would require us to multiply the first by 3 and the second by 2. This would produce a coefficient of 6 for x in both equations. An easier way would be to multiply the first equation by 3 and eliminate y from both equations, since this would involve only one multiplication. Thus,

$$(1) \quad 6x - 3y = 15$$
$$(2) \quad 3x + 3y = 21$$

from which $9x = 36$, $x = 4$, and (substituting the value for x in either of the original equations) $y = 3$.

Solve the following pairs of linear equations by use of the elimination procedure.

(a) (1) $3x - y = 21$ (c) (1) $7a + t = 42$
    (2) $2x + y = 4$ (2) $3a - t = 8$
(b) (1) $3k + 5p = 9$ (d) (1) $7r = 5s + 15$
    (2) $3k - p = -9$ (2) $2r = s + 9$

— — — — — — — — — — — — — —

(a) $x = 5$, $y = -6$; (b) $p = 3$, $k = -2$; (c) $a = 5$, $t = 7$; (d) $r = 10$, $s = 11$

14. As another example of the method of substitution for solving pairs of linear equations, consider the pair below:

$$(1) \quad 2x - y = 5$$
$$(2) \quad 3x + 3y = 21$$

Start by solving the first equation for y, that is, writing y as a function of x:

$$-y = 5 - 2x$$

or, multiplying both sides by -1 to change $-y$ to $+y$:

$$y = 2x - 5$$

Next, substitute the expression $2x - 5$ for y in the second equation, and solve for x:

$$3x + 3(2x - 5) = 21$$
$$3x + 6x - 15 = 21$$
$$9x = 21 + 15$$
$$9x = 36$$
$$x = 4$$

Substituting 4 for x in either of the original equations gives $y = 3$. Thus, we have the solution pair (4, 3).

Solve the following pairs of linear equations by the substitution procedure.

(a) (1) $y = 2x$
    (2) $7x - y = 35$

(c) (1) $a = b + 2$
    (2) $3a + 4b = 20$

(b) (1) $r = 4t - 1$
    (2) $6t + r = 79$

(d) (1) $3p = 27 - q$
    (2) $2q = 3p$

(a) $x = 7$, $y = 14$;   (b) $t = 8$, $r = 31$;   (c) $b = 2$, $a = 4$;   (d) $p = 6$, $q = 9$

UNIT NINE

Quadratic Equations

Review Item	Ref Page	Example
1. A *quadratic equation* in one unknown is an equation in which the highest power of the unknown is the second power.	160	Standard form: $ax^2 + bx + c = 0$ $3x^2 + 5x - 2 = 0$
2. A *complete quadratic equation* is one that contains first and second degree terms of the unknown as well as a nonzero constant term. An *incomplete quadratic equation* takes the form $ax^2 + c = 0$, which lacks the term containing the first power of the unknown, or the form $ax^2 + bx = 0$, where the constant term is zero.	161	Both equations cited in review item 1 are complete quadratic equations. $3x^2 - 12 = 0$ (no first power of the unknown) $2x^2 - 3x = 0$ (no constant term)
3. Many quadratic equations can be solved by the method for factoring a trinomial. (See review items 9 and 10, Unit 4.)	161	$x^2 - 7x + 12$ $= (x - 4)(x - 3)$ $4x^2 + 2x - 6$ $= 2(x - 1)(2x + 3)$

Review Item	Ref Page	Example
4. To solve a quadratic equation by factoring: • Write the quadratic equation in standard form ($ax^2 + bx + c = 0$). • Factor the left member into first degree factors. • Set each factor containing the variable equal to zero. • Solve the resulting first degree equations. • Check the roots of these first degree equations in the original quadratic equation.	162	Solve: $x^2 + 4x = 21$ Adding -21 to both members: $x^2 + 4x - 21 = 0$ $(x + 7)(x - 3)$ $x + 7 = 0; \; x - 3 = 0$ $x = -7; \; x = 3$; hence, -7 and 3 are the roots $(-7)^2 + 4(-7) \overset{?}{=} 21$ $49 - 28 \overset{?}{=} 21$ $21 \overset{\checkmark}{=} 21$ $3^2 + 4(3) \overset{?}{=} 21$ $9 + 12 \overset{?}{=} 21$ $21 \overset{\checkmark}{=} 21$
5. To solve an incomplete quadratic equation that lacks the constant term: • Factor the equation. • Set each term equal to zero. • Solve the resulting equations. • Check the roots in the original quadratic equation. One root will always be zero.	163	$x^2 - 2x = 0$ $x(x - 2) = 0$ $x = 0; \; x - 2 = 0$ M $\qquad x = 2$ Check: $0 - 0 = 0$ $4 - 4 = 0$
6. To solve an incomplete quadratic equation that lacks the first power of the unknown, write the equation so that the unknown term is on one side and the constant term on the other. Then use the method of extraction of roots.	164	$x^2 - 36 = 0$ $x^2 = 36$ $x = +6$

Review Item	Ref Page	Example
7. To transform a quadratic equation into standard form ($ax^2 + bx + c = 0$), use the axioms for transforming equations. These include: • Clearing the equation of fractions.	165	$x + \dfrac{7}{x} - 8 = 0$ becomes $x^2 - 8x + 7 = 0$
• Removing parentheses.		$x(x - 6) = -9$ becomes $x^2 - 6x + 9 = 0$
• Removing radical signs by squaring both members. • Collecting like terms.		$x^2 - 6x = 16$ becomes $x^2 - 6x - 16 = 0$ $3x^2 - 3 = 5x + 2x^2 - 7$ becomes $x^2 - 5x + 4 = 0$
8. To solve a quadratic equation by completing the square: • Eliminate the constant from the left member. • If the coefficient of the first (squared) term is not 1, divide both members by the coefficient. • Square half the coefficient of x and add to both sides.	165	$x^2 + 6x - 7 = 0$ $x^2 + 6x = 7$ (Not necessary in this case) $(\dfrac{1}{2} \cdot 6)^2 = (3)^2 = 9$ $x^2 + 6x + 9 = 7 + 9$
• Replace the trinomial by its binomial root squared. • Use the method of extracting roots. • Solve the resulting two equations. • Check both roots in the original equation.		$(x + 3)^2 = 16$ $x + 3 = +4$ $x + 3 = 4,\ x = 1;\ x + 3 = -4,$ $x = -7$

Review Item	Ref Page	Example
9. The trinomial in review item 8 *could* have been solved by factoring. To use the method of completing the square to solve a quadratic equation that cannot be solved by factoring, proceed as follows: • Eliminate the constant from the left member. • If the coefficient of the first (squared) term is not 1, divide both members by the coefficient. • Square half the coefficient of x and add to both sides. • Replace the trinomial by its binomial root squared. • Take the square root of both members of the equation. • Solve the resulting two equations. • Check both roots in the original equation.	166	$x^2 + 2x - 5 = 0$ $x^2 + 2x = 5$ (Not required in this case) $(\frac{2}{2})^2 = (1)^2 = 1$ $x^2 + 2x + 1 = 5 + 1$ $(x + 1)^2 = 6$ $x + 1 = +\sqrt{6}$ (1) $x = -1 + \sqrt{6}$ (2) $x = -1 - \sqrt{6}$ Substituting root (1); $(-1 + \sqrt{6})^2 + 2(-1 + \sqrt{6}) - 5 \overset{?}{=} 0$ $(1 - 2\sqrt{6} + 6) - 2 + 2\sqrt{6} - 5 \overset{?}{=} 0$ $7 - 2\sqrt{6} - 7 + 2\sqrt{6} \overset{?}{=} 0$ $0 \overset{\checkmark}{=} 0$ Check root (2) to prove it is equally valid.
10. Another method for solving quadratic equations is by use of the *quadratic formula*, which is a general solution of	166	General solution: $x = \dfrac{-b + \sqrt{b^2 - 4ac}}{2a}$

Review Item	Ref Page	Example
the standard form of the quadratic equation ($ax^2 + bx + c = 0$).		In the equation $x^2 - 5x + 6 = 0$, $a = 1$, $b = -5$, $c = 6$. Therefore, $$x = \frac{-(-5) + \sqrt{(-5)^2 - 4 \cdot 1 \cdot 6}}{2 \cdot 1}$$ $$x = \frac{5 + \sqrt{25 - 24}}{2}$$ $$x = \frac{5 + 1}{2} = 3, 2$$
11. To solve a quadratic equation by graphing: • Set the equation equal to y • Prepare a table of x and y values by assigning successive values to x and finding the corresponding values of y. • Plot the resulting curve.	167	$x^2 - x - 2 = 0$ $y = x^2 - x - 2$ <table><tr><td>x</td><td>3</td><td>2</td><td>1</td><td>0</td><td>-1</td><td>-2</td></tr><tr><td>y</td><td>4</td><td>0</td><td>-2</td><td>-2</td><td>0</td><td>4</td></tr></table>
12. To analyze the results of the graphic solution of an equation, observe the x intercepts, that is, the two points where the curve crosses the X axis. The values of x at these points are the roots of the	169	In the graphic solution for review item 11, $x = 2$, $x = -1$. (See example above)

Review Item	Ref Page	Example
equation. Note: • Two points of intersection indicate two unequal, real roots. • A point of tangency indicates a real root with a multiplicity of 2. • No intersection means no real roots.		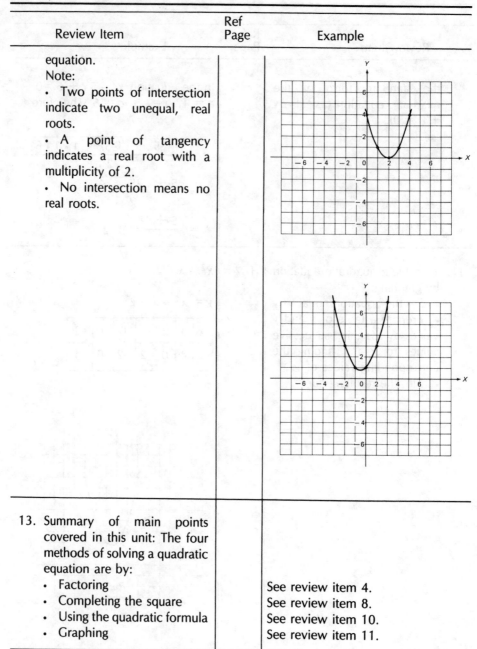
13. Summary of main points covered in this unit: The four methods of solving a quadratic equation are by: • Factoring • Completing the square • Using the quadratic formula • Graphing		See review item 4. See review item 8. See review item 10. See review item 11.

UNIT NINE REFERENCES

1. In the *standard form of a quadratic equation,* $ax^2 + bx + c = 0$, a represents the numerical coefficient of x^2 (the second degree term of the unknown); b is the numerical coefficient of x (the first degree term of the unknown); and c represents the constant term. In other words, a, b, and c represent the real number coefficients, and x represents the variable. In the equation $x^2 - 3x + 2 = 0$, $a = 1$, $b = -3$, and $c = 2$.

What are the values of a, b, and c in the equation $2x^2 - 49 = 0$?

- - - - - - - - - - - - - - -

$a = 2$; $b = 0$; $c = 49$

2. Quadratic equations are divided into two classes: *complete* and *incomplete quadratic equations*. A complete quadratic equation contains all three terms; an incomplete quadratic equation has one term missing. For example, in the equation $x^2 - 16 = 0$, the x term is missing in the equation; $x^2 - 6x = 0$, the constant term is missing.

Indicate which of the following are complete and which are incomplete.

(a) $x^2 + x - 2 = 0$ _____ (c) $3x^2 = 2x$ _____
(b) $2x - 7 = x^2$ _____ (d) $4x^2 - 49 = 0$ _____

- - - - - - - - - - - - - - -

(a) complete; (b) complete (but needs rearranging); (c) incomplete (lacks constant); (d) incomplete (lacks x term)

3. Although there are procedures to assist you in factoring a trinomial (as discussed in Unit 4), factoring some types of trinomials is a matter of educated guessing.

Consider the trinomial $x^2 + 2x - 15$. Since the numerical coefficient of x^2 is 1, the first term in each of the two binomial factors will simply be x. The next step is to examine the factors of 15 to discover if any two of them differ by 2 (the numerical coefficient of the middle term of the trinomial). Since 3 times 5 and 1 times 15 are the only integral factors of 15, it is evident that the correct pair of factors is 3 and 5.

It is now necessary to make the sign of the larger factor plus (+5) and the sign of the smaller factor minus (– 3) in order to arrive at the correct sign for the middle term. Thus, we get the binomial factors $(x + 5)(x - 3)$.

In a trinomial such as $4x^2 + 2x - 6$, we first have to use the distributive law to remove the common factor 2, giving us $2x^2 + x - 3$. Using the procedure given in Unit 4, we can factor this expression $(x - 1)(2x + 3)$, for a complete answer of $2(x - 1)(2x + 3)$.

For some further practice, factor the following quadratic polynomials. This will be helpful to you in solving quadratic equations, as we will be doing shortly.

(a) $x^2 + 4x - 21 =$ _____

(b) $x^2 + 6x + 8 =$ _____

(c) $4x^2 - 7x - 2 =$ _____

(d) $x^2 - 9 =$ _____

(e) $9x^2 - 6x + 1 =$ _____

(f) $3x^2 - x =$ _____

(g) $2y^2 - 5y - 25 =$ _____

(h) $k^2 - 4k =$ _____

- - - - - - - - - - - - - - -

(a) $(x + 7)(x - 3)$;   (b) $(x + 2)(x + 4)$;   (c) $(4x + 1)(x - 2)$;   (d) $(x - 3)$
$(x + 3)$; $(3x - 1)(3x - 1)$;   (f) $x(3x - 1)$; (g) $(y - 5)(2y + 5)$;   (h) $k(k - 4)$

Did you remember to classify problem (d) as the product of the sum and difference of the same two terms? Keep in mind also that problem (e) is the square
of a binomial. You should try to recognize and mentally identify such quadratic
expressions; they have special properties that can be used to advantage when
factoring.

4.   Having reviewed the procedure for factoring quadratic expressions, we can
consider how to solve equations involving quadratic expressions. There are
several methods available. The first one we will review is that of *solution by
factoring*. The method is outlined in review item 4. Here is another example.

Example: Solve: $x^2 - 6 = x$

Solution: Rewrite in standard form: $x^2 - x - 6 = 0$
         Factor left member: $(x + 2)(x - 3) = 0$
         Set each factor equal to zero: $x + 2 = 0; \; x - 3 = 0$
         Solve the resulting first
         degree equations: $x = -2; \; x = 3$ (hence -2 and 3
                                           are the roots)
         Check: $(-2)^2 - 6 \overset{?}{=} -2; \; 4 - 6 \overset{?}{=} -2$
                                           $-2 \overset{\checkmark}{=} -2$
                                           $(3)^2 - 6 \overset{?}{=} 3; \; 9 - 6 \overset{?}{=} 3$
                                           $3 \overset{\checkmark}{=} 3$

Note that the third step of this procedure uses the concept called the *zero
factor law*. This law states that if $a \cdot b = 0$, then $a = 0$, $b = 0$, or both a and
$b = 0$. This concept applies only when one member of an equation is zero.

Solve the following quadratic equations by the factoring method. Check each
of the roots in the original equation from which it was derived. You will not
know for certain whether or not you have made a mistake in deriving a numerical value as a root unless you test it in the original equation.

(a) $x^2 + 6x + 8 = 0$ (e) $3x^2 = x$

(b) $4x^2 - 7x - 2 = 0$ (f) $2y^2 - 5y = 25$

(c) $x^2 - 9 = 0$ (g) $k^2 - 4k = 0$

(d) $9x^2 - 6x = -1$

------ ---------

(a) $x = -2$; check: $4 - 12 + 8 \overset{?}{=} 0$, $0 \overset{\checkmark}{=} 0$

     $x = -4$; check: $16 - 24 + 8 \overset{?}{=} 0$, $0 \overset{\checkmark}{=} 0$

(b) $x = -\dfrac{1}{4}$, $x = 2$

(c) $x = 3$; check: $9 - 9 \overset{?}{=} 0$, $0 \overset{\checkmark}{=} 0$

     $x = -3$; check: $9 - 9 \overset{?}{=} 0$, $0 \overset{\checkmark}{=} 0$

(d) $x = \dfrac{1}{3}$, $x = \dfrac{1}{3}$ (this is an example of a multiple root)

(e) $x = 0$; check: $3(0)^2 \overset{?}{=} 0$, $0 \overset{\checkmark}{=} 0$

     $x = \dfrac{1}{3}$; check: $3(\dfrac{1}{3})^2 = \dfrac{1}{3}$, $\dfrac{1}{3} = \dfrac{1}{3}$

(f) $y = 5$, $y = -2\dfrac{1}{2}$

(g) $k = 0$; check: $0 - 0 \overset{?}{=} 0$, $0 \overset{\checkmark}{=} 0$

     $k = 4$; check: $16 - 16 \overset{?}{=} 0$, $0 \overset{\checkmark}{=} 0$

5. It is apparent that the method of solution by factoring will not always work, since not all quadratic equations are factorable. Now we review the procedure for solving incomplete quadratic equations in which the constant term equals zero. As indicated in review item 2, such equations take the form $ax^2 + bx = 0$.

If we factor the left member of such an equation, we get $x(ax + b) = 0$, from which $x = 0$ and $ax + b = 0$, representing two linear equations. If we solve the second equation for x, we get $x = -\dfrac{b}{a}$; hence, the roots of such an equation are zero and $-\dfrac{b}{a}$.

Example: Find the roots of the equation $y^2 - 2y = 0$.

Solution: We can solve this equation by factoring, which gives us $y(y - 2) = 0$, from which $y = 0$, and $y - 2 = 0$, $y = 2$. On the other hand, from our general solution above, we can simply write at once the second solution as

$$y = -\dfrac{-2}{1} = 2 \text{ (since in this case } a = 1 \text{ and } b = -2).$$

Solve the following equations using either the factoring method or the method based on the general solution.

(a) $4x^2 = 28x$ (remember to change to standard form $ax^2 + bx = 0$)

(b) $7k^2 - 35k = 0$

(c) $3a^2 - 18a = 0$

(d) $2x^2 + 3x = 0$

------ --------

(a) $x = 0, x = 7;$ (b) $k = 0, k = 5;$

(c) $3a^2 - 18a = 0;$

Factoring: $3a(a - 6) = 0;$

Two linear equations: $3a = 0, a - 6 = 0;$
                                            $a = 0, a = 6;$

(d) $2x^2 + 3x = 0;$
     $x(2x + 3) = 0;$
     $x = 0, 2x + 3 = 0;$
     $x = -\dfrac{3}{2}$ (i.e., $-\dfrac{b}{a}$)

6. Solving an incomplete quadratic equation of the type $ax^2 = c$ where the coefficient of the first degree term is zero, requires a somewhat different approach, since factoring may or may not work. If factoring does not work, we use the method of extraction of roots as summarized below:

Step 1: Solve for the square of the variable. This will yield an equation of the form $x^2 = \dfrac{c}{a}$.

Step 2: The roots of $x^2 = \dfrac{c}{a}$ will be the roots of the two equations $x = \sqrt{\dfrac{c}{a}}$ and $x = -\sqrt{\dfrac{c}{a}}$, if $\dfrac{c}{a}$ is positive. (There will be no real number solution if $\dfrac{c}{a}$ is negative.)

Step 3: Check results back in the original equation.

Example: Solve the equation $9x^2 - 25 = 0$ using the method of extraction of roots.

Solution: *Step 1:* $9x^2 = 25$, $x^2 = \dfrac{25}{9}$

Step 2: $x = \sqrt{\dfrac{25}{9}}$ and $x = -\sqrt{\dfrac{25}{9}}$, hence $x = \dfrac{5}{3}$ and $-\dfrac{5}{3}$, or $\pm\dfrac{5}{3}$

Step 3: Check: $9\left(\dfrac{5}{3}\right)^2 - 25 = 0$; $9\left(\dfrac{25}{9}\right) - 25 = 0$; $0 = 0.$

Solve the following equations by the method of extraction of roots. In problems (c) and (d), start by dividing both terms by the coefficient of x^2.

(a) $x^2 - 49 = 0$ (d) $3\ x^2 = 48$

(b) $x^2 = 36$ (e) $5\ x^2 = 16$

(c) $4x^2 - 36 = 0$

- - - - - - - - - - - - - -

(a) $x = \pm\ 7$;   (b) $= \pm\ 6$;   (c) $x = \pm\ 3$;   (d) $x = \pm\ 4$;   (e) $x = \dfrac{4\sqrt{5}}{5}$

7. A quadratic equation is not always in the standard form $ax^2 + bx + c = 0$. In fact, sometimes you will have to change the form of an equation just to see whether it *is* a quadratic equation. For example, $x = 4 - \dfrac{3}{x}$ may not at first appear to be a quadratic equation. However, clearing fractions and transforming the resulting equation gives us $x^2 - 4x + 3 = 0$, which is somewhat more recognizable.

If a quadratic equation is not in standard form, perform whatever operations are necessary to transform it to standard form. Any of the axioms we have developed for transforming to equivalent equations, such as those listed in review item 7, can be used.

Use any procedures necessary to express the following quadratic equations in standard form. Write the equation with a positive coefficient for the second degree term.

(a) $\sqrt{y^2 - 5y} = 3y$ (e) $3b^2 = -5b$

(b) $20 + 6k = 2k^2$ (f) $7(x^2 - 9) = x(x - 5)$

(c) $\dfrac{10}{c} + 1 = 4c$ (g) $18 = 2x^2$

(d) $p^2 = 5p - 4$ (h) $y(8 - 2y) = 6$

- - - - - - - - - - - - - -

(a) $8y^2 + 5y = 0$;   (b) $2k^2 - 6k - 20 = 0$;   (c) $4c^2 - c - 10 = 0$;   (d) $p^2 - 5p + 4 = 0$;

(e) $3b^2 + 5b = 0$;   (f) $6x^2 + 5x - 63 = 0$;   (g) $2x^2 - 18 = 0$;   (h) $2y^2 - 8y + 6 = 0$

8. We have mentioned that some quadratic equations cannot be solved by factoring. One method that can be used to solve such equations is called *completing the square*. This method requires that you make one member of the equation a perfect square. For example, the left member of the equation $x^2 + 6x - 7 = 0$ is not a perfect square trinomial because the third term, –7, does not have the correct value. However, the binomial $x^2 + 6x$ could be converted to a perfect square trinomial by adding a third term equal to the square of one-half the coefficient of x, that is, by adding 3^2 or 9. The necessary steps for doing this and working out the final solution of the equation are shown in review item 8.

Follow this procedure to solve these equations by completing the square. Do not forget to check your answers.

(a) $x^2 - 12x = 13$

(b) $a^2 - 6a = 7$

(c) $x^2 + 2x = 24$

(d) $m^2 - 12m + 11 = 0$

(e) $y^2 - 20y = 96$

(f) $k^2 + 18k + 17 = 0$

(g) $a^2 - 3a - 18 = 0$

(h) $2x^2 + 10x - 12 = 0$

(a) $x = 13, x = -1$; (b) $a = 7, a = -1$; (c) $x = 4, x = -6$;

(d) $m = 11, m = 1$; (e) $y = 24, y = -4$; (f) $k = -1, k = -17$;

(g) $a^2 - 3a - 18 = 0$

$a^2 - 3a + \dfrac{9}{4} = 18 + \dfrac{9}{4}$

$\left(a - \dfrac{3}{2}\right) = \dfrac{81}{4}$

$a - \dfrac{3}{2} = \pm\dfrac{9}{2}$

$a = 6, -3$

(h) $2x^2 + 10x - 12 = 0$

$x^2 + 5x + \dfrac{25}{4} = 6 + \dfrac{25}{4}$

$\left(x + \dfrac{5}{2}\right)^2 = \dfrac{49}{4}$

$x = 1, -6$

9. Use the method of completing the square to solve the following quadratic equations, which cannot be solved by factoring. Simplify all radicals.

(a) $x^2 - 6x + 2 = 0$

(b) $2x^2 + 4x - 8 = 0$

(c) $2x^2 - 3x - 1 = 0$

(a) $x = 3 \pm \sqrt{7}$; (b) $x = -1 \pm \sqrt{5}$; (c) $x = \dfrac{3 \pm \sqrt{17}}{4}$

10. From the method of completing the square, another method of solving quadratic equations can be derived, known as the *quadratic formula*. Its general form is as follows:

$$x = \frac{-b \pm \sqrt{b^2 - 4ac}}{2a}$$

If you plan to continue your study of mathematics, you will find it helpful to review the quadratic formula from time to time to make sure you remember it.

To help you gain (or regain) speed and ease in using it, here are some practice problems. Do not be concerned with whether or not they could be solved by factoring. Use the quadratic formula; then verify your results by factoring the equation, if possible. Be sure to check your answers by substituting them in the original equation. Before starting your work, write down the quadratic formula, so that you will have it handy for reference.

(a) $x^2 - 10x + 21 = 0$

(b) $x^2 + 2x - 3 = 0$

(c) $a^2 - 25 = 0$ (*Hint: b = 0*)

(d) $2k^2 - 22k + 60 = 0$ (*Hint:* Divide through by 2 first.)

(e) $y^2 - y - \dfrac{3}{4} = 0$

(f) $3x^2 - 5x - 2 = 0$ (*Hint: a = 3, b = -5, c = -2*)

- - - - - - - - - - - - - - -

(a) $x = \dfrac{-(-10) \pm \sqrt{(10)^2 - 4 \cdot 1 \cdot 21}}{2} = \dfrac{10 \pm 4}{2}$, or $x = 7, 3$;

(b) $x = \dfrac{-(2) \pm \sqrt{2^2 - (4)(1)(-3)}}{2} = \dfrac{-2 \pm 4}{2}$, or $x = 1, -3$;

(c) $a = \dfrac{0 \pm \sqrt{0 - (4)(1)(-25)}}{2} = \dfrac{\pm 10}{2}$, or $a = 5, -5$;

(d) $k = \dfrac{-(-11) \pm \sqrt{121 - (4)(1)(30)}}{2} = \dfrac{11 \pm 1}{2}$, or $k = 5, 6$;

(e) $y = \dfrac{1 \pm \sqrt{1 - (4)(1)(-\frac{3}{4})}}{2} = \dfrac{1 \pm 2}{2}$, or $y = \dfrac{3}{2}, -\dfrac{1}{2}$;

(f) $y = \dfrac{5 \pm \sqrt{25 - (4)(3)(-2)}}{6} = \dfrac{5 \pm 7}{6}$, or $x = 2, -\dfrac{1}{3}$

11. The last method we will discuss for solving quadratic equations is that of graphing.

In Unit 8, we reviewed the procedure for plotting a curve, as all plots of equations are termed, regardless of whether they are straight lines or curved. We found that, in plotting a curve on a rectangular coordinate system, we needed a pair of coordinates to locate any single point on the curve. These coordinates are the x coordinate (abscissa) and y coordinate (ordinate). The same general procedure for plotting a curve outlined in Unit 8 can be used to solve a quadratic equation.

Here is another example illustrating the procedure shown in review item 11:

Suppose we have the equation $x^2 - 2x - 3 = 0$ and wish to draw a graph in order to find its roots. The value of the left member, $x^2 - 2x - 3$, depends upon the value of x. Let us use ordered pairs to help us. As different values are assigned to x, the expression $x^2 - 2x - 3$ takes on different values. These may be thought of as values of y (the ordinate) for the corresponding values of x (the abscissa). We prepare a table of x and y values (just as you learned to do in Unit 8) assigning various values to x and finding the corresponding values of y.

If $x =$	4	3	2	1	0	–1	–2
Then y(or $x^2 - 2x - 3$) =	5	0	–3	–4	–3	0	5

Plotting these pairs of coordinates on graph paper we get the curve shown below.

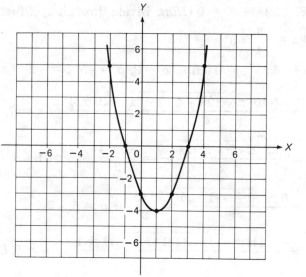

We know that:

- Our original equation was $x^2 - 2x - 3 = 0$.
- The solution or roots of the equation are those values of x that will make the equation a true statement (that is, will make the value of the left member become zero).
- Since $y = x^2 - 2x - 3$, the value of x that will make the expression $x^2 - 2x - 3$ zero will also make y zero.

Think carefully about these statements. Then see if you can tell from the graph what the roots are. The roots of the equation are $x =$ _____, and $x =$ _____.

3, –1.

Of course, you could have found the roots by factoring. I hope you did not fall back on the factoring method. What you should have observed is that, at the two points on the curve where $y = 0$, the curve crosses the X axis. Hence, *the roots of the equation are simply the values of x at these two points*. You could say, then, that the roots represent the values of the two points at which lines $y = 0$ and $y = x^2 - 2x - 3$ intersect. This is true of all quadratic equations that have real roots.

12.   The graph of a quadratic equation does not always cross the X axis. Sometimes it is tangent to it (that is, it touches the X axis at only one point). In that case, it will have two equal, real roots, more correctly expressed by saying that it will have *one root with a multiplicity of 2*. Observe this in the example below. The curve shown is tangent to the X axis. The roots are $x = 2, 2$.

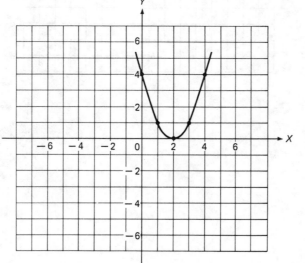

The other possibility is that the curve may not touch or intersect the X axis at all. In this case, there are no real roots. You will see this in the example below. The curve shown fails to intersect the X axis. There are no real roots. You will observe this in the example below.

The curve shown fails to intersect the X axis. There are no real roots.

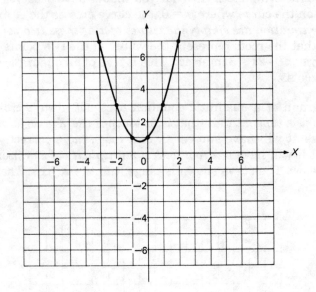

UNIT TEN

Inequalities

Review Item	Ref Page	Example
1. An *inequality* is a statement that the value of one expression is greater than or less than the value of another expression. A number to the *right* of another number on the horizontal number scale is the *greater* of the two numbers.	173	$5 > 2$ $3 < 7$ +-+-+-+-+-+-+-+-+-+-+-+ −5 −4 −3 −2 −1　0　+1 +2 +3 +4 +5 $+1 > 0$ $-2 > -3$ $0 > -4$
There are two principal inequality symbols, which mean *greater than* and *less than*.		$>$ means greater than $<$ means less than
2. In a *true statment of inequality*, the vertex of the inequality symbol always points to the *smaller* quantity, that is, the quantity farther to the left on the horizontal number scale.	174	$-2 > -4$ $1 > -10$ $-3 < 7$ $2 < 3$
For any two real numbers a and b, the statement $a > b$ is equivalent to the statement $a - b$ is positive.		If $a = 5$ and $b = 2$, then $5 > 2$, and $5 - 2 = +3$. If $a = -2$ and $b = -5$, then $-2 > -5$, and $-2 - (-5) = +3$.

Review Item	Ref Page	Example
3. An *absolute inequality* is one that is true for all values of the variables involved. A *conditional inequality* is one that is true only for certain values of the variables involved. An *inconsistent inequality* is one which is not a true statement for any value of the variables.	175	$x + 1 > x$ $y - 4 > 0$ (true only if y is greater than 4) $x^2 < 0$ (not true for any real number, because the square of any real number is greater than or equal to zero)
4. If the symbols of two inequalities point in the same direction, the inequalities have the *same sense*. If the symbols point in opposite directions, the inequalities have the *opposite sense*. The sense of an inequality is *unchanged* if the same number is added to or subtracted from both sides. The sense of an inequality is *unchanged* if both sides are multiplied or divided by the same *positive* number. The sense of an inequality is *changed* if both sides are multiplied or divided by the same *negative* number.	175	$a > b$ and $c > d$ (same sense) $a > b$ and $c < d$ (opposite sense) Since $5 > 4$, then $5 + 2 > 4 + 2$, or $5 - 2 > 4 - 2$. Since $5 > 4$, then $5 \cdot 2 > 4 \cdot 2$, and $\frac{5}{2} > \frac{4}{2}$. Since $5 > 4$, then $5(-1) < 4(-1)$, and $\frac{5}{-1} < \frac{4}{-1}$.
5. If the multiplier of an inequality is positive, the inequality sign stays the same. If the multiplier of an inequality is negative, the direction of the inequality symbol must be reversed.	177	$(3 < 6) \cdot (2)$ $6 < 12$ $(3 < 6) \cdot (-2)$ $-6 > -12$

Review Item	Ref Page	Example
6. Terms may be moved from one side of an inequality to the other side according to the rules that hold for equations.	177	If $6x - 4 > 3x + 2$, then $6x - 3x > 2 + 4$, $3x > 6$, and $x > 2$.
7. Equations and inequalities can be solved by both algebraic and graphic methods.	178	Solve: $x + 1 > 0$ Algebraic solution: subtract 1 from each side: $x > -1$ Graphic solution:
8. Linear inequalities in two variables can be solved graphically using a two dimensional rectangular coordinate system.	179	See reference item 8.

In row 7, Graphic solution:

$$-4 \quad -3 \quad -2 \quad -1 \quad 0 \quad +1 \quad +2 \quad +3$$

UNIT TEN REFERENCES

1. Equations and inequalities are either true statements or false statements; they cannot be both at the same time. One of our primary concerns with inequalities, as with equations, is to find the value or values that make a given statement a true statement. This constitutes solving the inequality.

In Unit 2, we reviewed the concept of the number scale or number line to help explain positive and negative numbers. Thus, in the horizontal number scale below, positive numbers are defined as those to the right of zero, and negative numbers as those to the left of zero.

$$-8 \quad -7 \quad -6 \quad -5 \quad -4 \quad -3 \quad -2 \quad -1 \quad 0 \quad +1 \quad +2 \quad +3 \quad +4 \quad +5 \quad +6 \quad +7 \quad +8 \quad +9 \quad +10$$

Any number to the right of another number on the number scale is the greater of the two numbers. The symbol $>$ means "greater than." Looking at the number line above, we can see that $+4 > +1$, meaning that positive 4 is greater than positive 1, and that $+2 > -6$, meaning that positive 2 is greater than negative 6.

Conversely, the symbol $<$ means "less than." Thus, the two relationships mentioned above can also be expressed as $+1 < +4$, meaning that positive 1 is less than positive 4, and – 6 < +2, meaning that negative 6 is less than positive 2.

Indicate whether the statements of inequality below are true or false. Use the number scale to assist you.

(a) +5 > +4 _____   (f)   0 < -2 _____

(b) +1 > 0 _____   (g) -6 < -8 _____

(c) -4 < -6 _____   (h)   4 > 0 _____

(d) -1 < +1 _____   (i)   9 < -10 _____

(e) -5 < -1 _____   (j) -1 > -8 _____

------ ---------

(a) true;   (b) true;   (c) false;   (d) true;   (e) true;   (f) false;   (g) false;
(h) true;   (i) false;   (j) true

2. If you look again at the sample inequalities above, you will see that in each case where the statement is a true statement, the vertex, or point, of the inequality symbol always points toward the smaller number. If this is not the case, then the statement is not true.

We can also make the following general statement:

> *For any two real numbers a and b, the statement a > b is equivalent to the statement a – b is positive. Similarly, the statement a < b is equivalent to the statement a – b is negative.*

Note that the expression $a - b$ must be either positive, zero, or negative; or to say the same thing another way, given any two real numbers, only one of the following is true:

$$a > b \qquad a = b \qquad a < b$$

Bearing in mind that the vertex of the inequality symbol always points to the smaller value, insert the correct symbol in the problems below.

(a) -3 ___-2   (d) +3 ___+4

(b) +1 ___-6   (e)   0 ___+1

(c) -1 ___+6   (f) -8 ___0

------ ---------

(a) – 3 < -2; (b) +1 > – 6; (c) –1 < +6; (d) +3 < +4; (e) 0 < +1; (f) – 8 < 0

3. Using the definitions given in review item 3, indicate whether each of the following inequalities is absolute, inconsistent, or conditional.

(a) $a^2 - b^2 > 0$ _____

(b) $a^2 > 0 \ (a \neq 0)$ _____

(c) $x^2 + 1 > 0$ _____

(d) $2x + 3 > 5x - 9$ _____

(e) $(x + 3)^2 < 0$ _____

- - - - - - - - - - - - - -

(a) conditional; (b) absolute; (c) absolute; (d) conditional; (e) inconsistent

4. As we discuss some of the properties of inequalities, you will get some feeling for the ways in which they can be handled, just as you have probably already developed a "feel" for equations. To begin with, note that when the signs of two inequalities point in the same direction, they have the *same sense,* and when they point in opposite directions they have the *opposite sense.*

This terminology helps us talk about some of the things that happen when the familiar operations of addition, subtraction, multiplication, and division are performed on inequalities. Let us review the properties stated in review item 4.

> *Property 1:* The sense of an inequality is unchanged if the same num-
> ber is added to or subtracted from both sides of the in-
> equality.

For example,

$$4 > 3$$

$$4 + 2 > 3 + 2$$

$$4 - 2 > 3 - 2$$

In general terms, if $a > b$, then $a + c > b + c$ for any number c.

> *Property 2:* The sense of an inequality is unchanged if both sides of the
> inequality are multiplied or divided by the same positive
> number.

For example,

$4 > 3$

$4 \cdot 2 > 3 \cdot 2$

$4 \div 2 > 3 \div 2$

In general terms, if $a > b$, then $a \cdot c > b \cdot c$ and $a \div c > b \div c$ if $c > 0$.

Property 3:   The sense of an inequality is changed if both sides of the inequality are multiplied or divided by the same negative number.

For example,

$4 > 3$

$4(-1) < 3(-1)$

$\dfrac{4}{-1} < \dfrac{3}{-1}$

In general terms, if $a > b$, then $a \cdot c < b \cdot c$ if $c < 0$.

Insert the correct inequality symbols in the following.

(a) $7 + 2$ _____ $5 + 2$

(b) $8 \cdot 3$ _____ $3 \cdot 3$

(c) $12 - 5$ _____ $6 - 5$

(d) $\dfrac{6}{2}$ _____ $\dfrac{4}{2}$

(e) $9(-2)$ _____ $4(-2)$

(f) $\dfrac{11}{-3}$ _____ $\dfrac{7}{-3}$

In some of the following problems, the original terms of the inequality have had the same amount added to or subtracted from each side; in others, they have been multiplied or divided by the same amount. Your job is to insert the correct inequality sumbol.

(g) If $4 < 7$, then $4 + 3$ _____ $7 + 3$

(h) If $-2 > -7$, then $-2 + 5$ _____ $-7 + 5$

(i) If $4 < 5$, then $4(-3)$ _____ $5(-3)$

(j) If $8 > 4$, then $8 - 2$ _____ $4 - 2$

(k) If $-4 < -2$, then $-4(3)$ _____ $-2(3)$

(l) If $9 > 7$, then $\dfrac{9}{3}$ _____ $\dfrac{7}{3}$

- - - - - - - - - - - - - - -

(a) $>$; (b) $>$; (c) $>$; (d) $>$; (e) $<$; (f) $<$; (g) $<$; (h) $>$; (i) $>$ (changes because multiplier is negative); (j) $>$; (k) $<$; (l) $>$

5. The second and third properties listed in review item 4 tell us what happens when both sides of an inequality are multiplied or divided by the same number: If the multiplier is positive, the inequality sign stays the same; if the multiplier is negative, the direction of the inequality symbol is reversed.

Because forgetting this is a common source of error in working with inequalities, you would do well to work a few more problems just to make sure you are aware of this point.

In the problems below, you are given an inequality and a multiplier. Multiply both sides of the inequality by the multiplier to arrive at a new inequality.

Original Inequality	Multiplier	New Inequality
(a) $8 > 7$	2	_____
(b) $2 > -1$	-3	_____
(c) $-11 < -5$	1	_____
(d) $5 < 7$	5	_____
(e) $12 < 17$	-1	_____
(f) $-8 < 7$	2	_____

- - - - - - - - - - - - - - -

(a) $16 > 14$;   (b) $-6 < 3$;   (c) $-11 < -5$;   (d) $25 < 35$;   (e) $-12 > -17$; (f) $-16 < 14$

6. The three properties discussed in review item 4 allow us to work with inequalities in the same way we work with equations. We can use these properties to change the form of the inequality until we can see the value or range of values of the variable that will make the statement of inequality a true statement.

Example 1:
Solve for x: $\qquad\qquad\qquad\qquad\qquad\qquad\qquad 4x - 6 > 2x + 2$

Solution:

Subtracting $2x$ from and adding 6 to both members: $\qquad\qquad\qquad 4x - 2x > 2 + 6$

Simplifying both members: $\qquad\qquad\qquad\qquad\qquad 2x > 8$

Dividing both members by positive 2: $\qquad\qquad\qquad\quad x > 4$

The solution tells us that the original statement will be true if the variable x is replaced by any number greater than positive 4. To check this, we select any number greater than 4 and replace the variable with that value. If we select the number 5, for example, we would get 14 in the left member and 12 in the right. Since $14 > 12$, the statement is true.

Example 2:   To emphasize the similarities in the way equations and inequalities can be solved, compare the following two solutions:

Equation	Inequality
$x - 5 = 3$	$x - 5 > 3$
$(x - 5) + 5 = 3 + 5$	$(x - 5) + 5 > 3 + 5$
$x = 8$	$x > 8$

Work the following problems in the same step-by-step manner shown in Example 1 above.

(a) $7x - 5 > 3x + 4$ (f) $3x + 4 > 2x - 1$

(b) $9x + 2 < 4x - 3$ (g) $4k - 2 < k + 7$

(c) $5 + 4y > 2y + 1$ (h) $2x + 6 < 7x - 4$

(d) $6 + 2y < 5y + 9$ (i) $-3 + m > 7 - 2m$

(e) $2y - 7 < y + 3$

- - - - - - - - - - - - - -

(a) $x > \dfrac{9}{4}$; (b) $x < -1$; (c) $y > -2$; (d) $y > -1$; (e) $y < 10$;

(f) $x > -5$; (g) $k < 3$; (h) $x > 2$ (direction of inequality symbol changes because both sides are divided by a negative number); (i) $m > 3\dfrac{1}{3}$

7.   In Unit 9, you saw that equations can be solved by both algebraic and graphic methods. The same is true of inequalities. We have looked at some algebraic solutions for inequalities. It is time now to talk about the graphic methods of solution.

Think for a moment about the inequality $x > 3$. What does it say? It says that the solution values of x can be selected from the entire set of numbers greater than 3. Can you see how we might indicate this idea on a horizontal number scale? It should indicate clearly that all numbers greater than 3 are included, while all numbers less than or equal to 3 are excluded. One conventional way to represent it is like this:

There are several ways of indicating that the number 3 itself is not included. We have used the hollow circle placed at the number to show this.

However, in some inequalities, it might be necessary to include the number 3 as a possible value. There is a special symbol to represent this: $\geqslant$, which means "greater than or equal to." If the inequality had been $x \geqslant 3$, we would have shown that the 3 was included by filling in the circle above it. Thus the graph of $x \geqslant 3$ would look like this:

As you might infer, there is a fourth common inequality symbol: $\leqslant$, which means "less than or equal to."

Draw a graph on a horizontal number scale to represent each of these inequalities. In each inequality, n represents some number.

(a) $n > 3$

(b) $n < -1$

(c) $n \geqslant -2$

(d) $n \leqslant 4$

- - - - - - - - - - - - - -

(a)

(b)

(c)

(d)

8. The simple concepts we have discussed so far form the basis for further work with inequalities containing two variables rather than just one. The number-line graphs we have used to plot inequalities in one variable will not serve for plotting inequalities in two variables. We will use instead the two-dimensional rectangular coordinate system we worked with in Units 8 and 9 when plotting linear and quadratic equations.

Consider, for example, the inequality $2x - y + 4 > 0$. This symbolic statement is telling us that there exists an entire group of points (x, y) such that the value of the expression $2x - y + 4$ is greater than zero. If we set the expression equal to zero (that is, $2x - y + 4 = 0$), we will get a linear equation that will plot as a straight line on one side of which lie all the points giving a value greater than zero, and on the other side all the points giving a value less than zero.

To graph this equation, we will find the x and y intercepts together with one additional check point, plot these three points, then draw the straight line.

Given:

$$2x - y + 4 > 0$$

$$2x - y + 4 = 0$$

$$y = 2x + 4$$

x	0	-2	-1
y	4	0	2

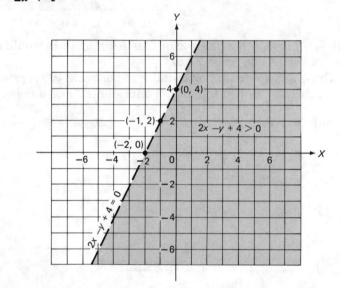

All that remains is to find out on which side of the line the values lie that make the inequality a true statement; that is, $2x - y + 4$ is *greater* than zero. Many points could be used to test this inequality, but the easiest point probably would be $(0, 0)$, where the two axes cross. Substituting these values in the inequality, we get $2(0) - 0 + 4 > 0$, or $4 > 0$. This point is included in the coordinate half-plane (that is, half of the total coordinate plane) lying to the right of the line. Therefore, we can conclude that all the points lying to the right of the line are greater than zero (positive) and, by inference, that all the points lying in the left half-plane are less than zero (negative). This is why the right half-plane is shaded in the graph above, indicating that this is the area containing all the points whose values make the inequality a true statement.

Use the same general procedure to graph the following inequalities and to determine in which half-plane they are true.

(a) $2x + y - 3 > 0$

(b) $2x - y - 2 > 0$

(c) $2x - 5y < 0$

- - - - - - - - - - - - - -

(a) $2x + y - 3 > 0$
$2x + y - 3 = 0$
$y = 3 - 2x$

At origin (0, 0) we get:

$2(0) + 0 - 3 \overset{?}{>} 0$
$-3 \overset{?}{>} 0$
$-3 \not> 0$

Therefore, $2x + y - 3 > 0$ is true only in the right half-plane.

x	0	$\frac{3}{2}$	1
y	3	0	1

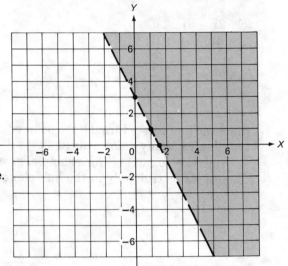

(b) $2x - y - 2 > 0$
$2x - y - 2 = 0$
$y = 2x - 2$

At origin (0, 0) the inequality becomes:

$2(0) - 0 - 2 \overset{?}{>} 0$
$-2 \not> 0$

Therefore, $2x - y - 2 > 0$ is true only in the right half-plane.

x	0	1	3
y	-2	0	4

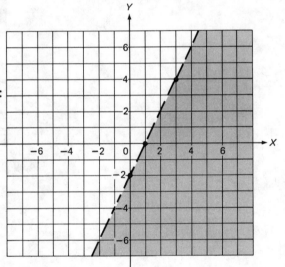

(c) $2x - 5y < 0$
$2x - 5y = 0$
$$x = \frac{5y}{2}$$

Since the line passes through the point or origin, you could try substituting the coordinates of any point near (0, 0).

Using the point (1, 1) for example, gives us $2(1)-5(1) \overset{?}{<} 0$, or $-3 \overset{\checkmark}{<} 0$. Hence $2x - 5y < 0$ is true only in the half-plane to the left of the graph line.

x	0	$\frac{5}{2}$	5
y	0	1	2

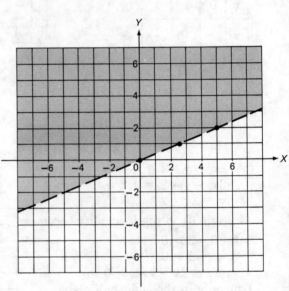

UNIT ELEVEN
Ratio, Proportion, and Variation

Review Item	Ref Page	Example
1. *Ratio:* If *a* is a quantity expressed in some unit of measure and *b* is some quantity expressed in the same unit of measure, then the ratio of *a* to *b* is the quotient $\frac{a}{b}$ (or *a/b*). Alternate forms of the ratio are *a:b* and *a ÷ b*.	186	The ratio of 3 days to 7 days is $\frac{3}{7}$. The ratio of 2 pounds to 17 ounces is 32:17. The ratio of 3 hours to 7 gallons is not defined.
2. *Proportion:* A statement that one ratio is equal to another ratio.	187	2:3 = 4:6 If *a* : *b* and *c* : *d* are equal ratios, then *a* : *b* = *c* : *d* is a proportion.
3. *Terms* of a proportion: The members of a proportion.	188	In the proportion 3:7 = 9:21, 3, 7, 9, and 21 are, respectively, the first, second, third, and fourth terms.
Means of a proportion: The middle terms of a proportion. *Extremes* of a proportion: The terms of a proportion that are farthest apart.		In the proportion 3:7 = 9:21, 7 and 9 are the means, and 3 and 21 are the extremes.

Review Item	Ref Page	Example
n any proportion, the product of the means is equal to the product of the extremes.	188	In the proportion $3:4 = 6:8$, $3 \cdot 8 = 4 \cdot 6$; and if $a : b = c : d$, then $a \cdot d = b \cdot c$.
5. *Interchanging the extremes and/or means:* If two ratios form a proportion, the ratios formed by interchanging the extremes and/or the means will also be proportional.	188	Since $2:3 = 4:6$ is a proportion, $6:3 = 4:2$ is a proportion by interchanging the extremes. Since $3:5 = 12:20$ is a proportion, $3:12 = 5:20$ is a proportion by interchanging the means.
6. *Direct proportion:* If the components a and b are related in such a way that every quotient $\dfrac{a}{b}$ has the same value, then a is directly proportional to b.	189	The cost of several cans of fruit is directly proportional to the cost of one can of fruit (assuming that no quantity discount is involved). If a is always twice as large as b, then the ratio of $\dfrac{a}{b}$ is $\dfrac{2}{1}$ for each pair (a, b), and a is directly proportional to b.
7. *Inverse proportion:* If the components a and b are related in such a way that every product $a \cdot b$ has the same value, then a is inversely proportional to b.	190	The rate of speed needed to cover a certain distance is inversely proportional to the time spent covering the distance. If $a \cdot b = 12$ for each pair of values given the pair (a, b), then a is inversely proportional to b.

Review Item	Ref Page	Example
8. *Variation* is another method of showing the relation between two variables.	192	The equations $\frac{x}{y} = k$, $x \cdot y = k$, $\frac{x}{yz} = k$, and $\frac{xy}{z} = k$ are examples of variation.
Direct variation: If two variables *a* and *b* are related in such a way that the quotient *a*/*b* has the same value for any ordered pair (*a*, *b*), then *a* is said to vary directly as *b*.		If $\frac{a}{b} = 6$ for all values of *a* and *b*, then *a* varies directly as *b*, and 6 is the constant of variation (*k*).
9. *Inverse variation:* If two variables *a* and *b* are related in such a way that the product *a* • *b* has the same value for all ordered pairs (*a*, *b*), *a* is said to vary inversely as *b*.	193	If $a \cdot b = 12$ for all values of *a* and *b*, then *a* varies inversely as *b*, and 12 is the constant of variation (*k*).
10. *Joint variation:* If *x* varies directly as the product of two or more variables, then *x* varies jointly as these variables.	194	If *x* varies as the product *abc*, then *x* varies jointly as *a*, *b*, and *c*. This can be represented by the equation $\frac{x}{abc} = k$
Combined variation: If *x* varies directly as one or more variables and inversely as a second or other variable, it is said to be a combined variation.		If *x* varies directly as *z* and inversely as *y*, the three quantities are an example of combined variation. This can be represented by the equation $\frac{xy}{z} = k$

Review Item	Ref Page	Example
11. Once the constant of variation has been found, the next step in solving a variation problem is to express the equation of variation in the most convenient form.	195	For direct variation: $\dfrac{a}{b} = k$; or $\dfrac{x}{y} = 3$, $x = 3y$ For inverse variation: $a \cdot b = k$; or $xy = 3$, $x = \dfrac{3}{y}$
12. The final step in solving a variation problem is to evaluate the equation, that is, to find the value of the unknown variable when the values of the other variables are known.	196	Substituting the value $y = 4$ in each of the equations above we get: $x = 3y = 3 \cdot 4 = 12$; $x = \dfrac{3}{y} = \dfrac{3}{4}$

UNIT ELEVEN REFERENCES

1.   One method of comparing quantities is to find their difference. Thus, if one person is 22 years of age and another 33, the difference in their ages is 11. So we can say that the first person is 11 years younger than the second, or that the second is 11 years older than the first.

A second method of comparing these two ages is to say that one person's age is $\dfrac{2}{3}$ that of the other. When we compare by this method, we are using the concept of *ratio*. We are saying, in effect, that for each 2 years of the younger person's life, the older person has had 3 years of life. Using symbols, we can write the ratio of their ages as $\dfrac{2}{3}$, 2/3, 2:3, or 2 ÷ 3.

The ratio of two like quantities is defined as the quotient of the first quantity divided by the second. This quotient usually is expressed in lowest terms. Thus, the ratio of 20 to 50 is written as 2:5.

As review item 1 specifies, the two quantities being compared must be *like quantities,* that is, that they be in the same units of measure. Sometimes this requires us to convert them into the same units before we can form a correct ratio. For example, the ratio of 3 days to 1 week is found by changing 1 week into 7 days and writing the ratio as 3:7.

We have defined ratio as the quotient of two *like* quantities. However, you will often see a ratio expressed between two quantities that are entirely different in nature. For example, in reference item 15 of Unit 7, we expressed speed (S) as the ratio of distance (D) to time (T):

$$S = \frac{D}{T}$$

If a and b do not represent quantities of the same kind, the ratio $a:b$ simply represents the portion of a that corresponds to one unit of b, as 1 mile per hour.
Write each of the following ratios in simplest form.

(a) 7 dozen to 6 dozen

(b) 3 yards to 7 feet

(c) 85 pounds per square inch to 150 pounds per square inch

(d) 36 square feet to 3 square yards

(e) 25 gallons to 50 gallons

(f) 15 days to 36 quarts

(g) 12 pounds to 36 yards

(h) 45 dimes to 12 quarters

(i) 18°F to 30°F (F indicates Fahrenheit scale)

(j) 3 cubic feet to 1 cubic yard

------ ---------

(a) 7:6;   (b) 9:7;   (c) 85:150 or 17:30;   (d) 36:27 or 4:3;   (e) 25:50 or 1:2;
(f) cannot be done, units not compatible;   (g) same as (f);   (h) 45:30 or 3:2;
(i) 18:30 or 3:5;   (j) 3:27 or 1:9

2. You will be using your ability to form and use ratios as you begin to work with proportions. This is because, as review item 2 states, a proportion is merely a statement that one ratio is equal to another ratio. Thus, $\frac{1}{2} = \frac{3}{6}$, or $1:2 = 3:6$, is a proportion.

Which of the following pairs of ratios could be used to form a true proportion? (*Remember:* The two ratios must *equal* one another.)

(a) $\frac{2}{3}$ and $\frac{3}{4}$ (c) $\frac{2}{7}$ and $\frac{6}{21}$

(b) $\frac{3}{4}$ and $\frac{6}{8}$ (d) $\frac{3}{7}$ and $\frac{5}{9}$

- - - - - - - - - - - - -

(b) and (c)

3. It will be helpful to learn a few special names and some important properties of proportions. In the proportion $a:b = c:d$ (which we usually read "a is to b as c is to d"), the letters a, b, c, and d are called, respectively, the first, second, third, and fourth terms. The terms a and d are called the *extremes*, since they are farthest apart, and b and c are called the *means*, since they are the middle terms.

In the proportion $5:7 = 10:14$, the extremes are _____ and _____, and the means are _____ and _____.

- - - - - - - - - - - - -

5, 14, 7, 10

4. Answer the following questions based on the fact that in any proportion the product of the means is equal to the product of the extremes.

(a) In the proportion $2:7 = 3:x$, the product of the extremes is _____ and the product of the means is _____.

(b) If $3:x = 4:7$ is a proportion, then $4x =$ _____.

(c) If $n:a = c:r$ is a proportion, then _____ $= nr$.

- - - - - - - - - - - - -

(a) $2x$ and 21;   (b) 21;   (c) ac

5. As stated in review item 5, the positions of the extremes and/or means of a proportion may be interchanged. This can be expressed in symbols as follows:

If $a:b = c:d$, then $a:c = b:d$, $d:b = c:a$, and $d:c = b:a$

Examples: If $2:3 = x:y$, then $2:x = 3:y$.
If $2:5 = x:y$, then $5:2 = y:x$.
If $3:4 = 6:8$, then $8:4 = 6:3$.

Following is an example of the use of direct proportion to solve a practical problem.

Example: If a certain parking lot contains Fords and Plymouths in the ratio 5:3, and there are 900 Plymouths in the lot, how many Fords are there in the lot?

Solution: Let x equal the number of Fords in the lot.
              Then: $5 : 3 = x : 900$
                     Multiplying the means and extremes : $3x = 4500$
                     $x = 1500$ (the number of Fords)

Solve the following problems using the idea of direct proportion.

(a) An aircraft has a wing span of 130 feet. The height from the ground to the top of the tail is 25 feet. If on a drawing the wing span is 42 inches, how many inches will be needed to show the distance from the ground to the top of the tail?

(b) If an automatic billing machine can process 21,000 cards in 14 hours, how many cards could be processed in one 8-hour shift?

(c) The vibrating frequency of a strong is directly proportional to the square root of the tension on the string. If a bass violin string vibrates 254 times per second when the tension is 5 pounds, what would the vibrating frequency be when the tension is 20 pounds?

(d) The distance that a falling body drops is directly proportional to the square of the time it has fallen. If a body falls 64 feet in 2 seconds, how far will it fall in 10 seconds?

— — — — — — — — — — — — — — —

(a) $130{:}42 = 25{:}x$; $x = 8\dfrac{1}{13}$ inches;

(b) $21{,}000{:}14 = x{:}8$; $= 12{,}000$ cards;

(c) $254{:}5 = x{:}20$; $x = 508$ times per second;

(d) $64{:}(2)^2 = x{:}(10)^2$; $x = 16000$ feet

Note: Although the two numbers in a ratio must be in the same units, the two ratios comprising a proportion can be, and usually are, in different units, as in the previous problems.

6. Now that we have defined the concept of proportion and discussed some of its properties, it is time to illustrate some of the various types of proportions. One kind of proportion is the *direct proportion,* illustrated in the following problem:

Suppose you want to make a table of distances covered (d) according to time traveled (t) for someone traveling at a constant rate (r) of 60 miles per hour. Using the formula $d = rt$, you could develop the following table:

t = elapsed time in hours	1	2	3	6	10	12
d = distance in miles	60	120	180	360	600	720

Notice that the ratio of time to distance is always 1:60 (read "one to sixty"). We can write this as a continued proportion:

$$\frac{t}{d} = \frac{1}{60} = \frac{2}{120} = \frac{3}{180} = \cdots \frac{12}{720}$$

Any pair of ratios from this table can be used to form a proportion. When the quotient of any two quantities (such as the ordered pairs in the table above) is constant, we have a *direct proportion* between the two quantities.

In most proportion problems, three terms are known; you are required to find the missing term.

Example: Find x if: $3:x = 9:18$

Solution   Products of the means and extremes: $9x = 54$
             From which: $x = 6$

Use the fact that if $a:b = c:d$ then $a \cdot d = b \cdot c$ to find the missing term in each of the following proportions.

(a) $3:4 = x:8$ (f) $x:5 = 5:x$

(b) $a:7 = 7:3$ (g) $3a:5 = 6:1$

(c) $(a-3):6 = 3:12$ (h) $3:(x-2) = x:1$

(d) $5:(x-2) = 4:7$ (i) $3:(x-1) = x:2$

(e) $3:x = x:12$ (j) $4:3 = x:(x+1)$

------ ---------

(a) $x = 6$;   (b) $a = 16\frac{1}{3}$;   (c) $a = 4\frac{1}{2}$;   (d) $x = 10\frac{3}{4}$;   (e) $x = \pm 6$;
(f) $x = \pm 5$;   (g) $a = 10$;   (h) $x = 3, -1$;   (i) $x = 3, -2$;   (j) $x = -4$

7.   The counterpart of direct proportion is *inverse proportion*. In an inverse proportion, the product of the numbers in one ordered pair is the same as the product of the numbers in the other ordered pair. Using symbols, we would represent this as $a \cdot b = c \cdot d$. A slightly more formal definition is as follows:

>   *If a set of ordered pairs (a, b) has the property that a is inversely proportional to b, then a · b is the same value for all of the ordered pairs (a,*

b). Or, using symbols, if a • b = k for any of the ordered pairs (a, b), then a is inversely proportional to b.

The value of k is called the *constant of of proportionality.* The table below, showing the relationship between pressure and volume of a gas, illustrates how inverse proportion differs from direct proportion. (For the purpose of this example, we assume that the container holding the gas can expand and that the temperature of the gas remains constant.)

p = pressure in pounds per square inch	50	40	20	15	10
v = volume in cubic inches	200	250	500	$66\frac{2}{3}$	1000

The product of the ordered pairs in this table is constant; it is always 10,000.
  The table below is another example of an inverse proportion.

a	6	4	12	24	
b	8	12	4		-12

Let us use this table to practice setting up some proportions.

Example 1: Set up a proportion and find b when a has a value of 24.

Solution: Some possible proportions are:

$$\frac{6}{24} = \frac{b}{8}, \quad \frac{6}{b} = \frac{24}{8}, \quad \frac{4}{24} = \frac{b}{12}, \quad \frac{12}{24} = \frac{b}{4}$$

Many other proportions could be established by selecting various ordered pairs (a, b) and/or interchanging the extremes or means. In any of the proportions, however, the value of b will be 2 when the value of a is 24. When setting up inverse proportions, remember that since $a • b = c • d$, then $\frac{a}{c} = \frac{d}{b}$.

Example 2: Set up a proportion and find the value of a when b is –12.

Solution: The proportion could be $\frac{6}{a} = \frac{-12}{8}$ or some other proportion obtained from the table or ordered pairs and using the definition of inverse proportion. When b is –12, the value of a is –4.

With the above examples as a general guide, try setting up proportions to solve the following problems.

(a) If x is inversely proportional to y and $x = 6$ when $y = 12$, find the value of x when $y = 4$. Show the proportion and the solution. *Hint:* Remember that if x is inversely proportional to y, then the product of every (x, y) pair is a constant (that is, the same). Thus, in this problem we know that $x • y = 6 • 12 = x • 4$ and we can use any pair of these equal ratios to set up a proportion.

(b) If r is inversely proportional to d^2 and $r = 3$ when $d = 5$, find r if $d = 8$.

(c) If y is inversely proportional to $4x + 5$ and $y = 4$ when $x = 8\frac{1}{2}$, find x when $y = 12$.

(d) The intensity of light (i) on a surface is inversely proportional to the square of the distance (d) between the surface and the light source. A light is 3 feet from a surface which receives a certain amount of light. What distance will cause the surface to receive ¼ the present amount of light? *Hint:* What is true about the pairs of values (i, d^2)? Also, remember that $\dfrac{i}{\frac{1}{4}i} = \dfrac{1}{\frac{1}{4}}$.

- - - - - - - - - - - - - - - -

(a) The proportion should be some form of $\dfrac{6}{4} = \dfrac{x}{12}$; $x = 18$ is the answer.

(b) The proportion should be some form of $\dfrac{3}{8^2} = \dfrac{r}{5^2}$; the answer is $r = \dfrac{75}{64}$.

(c) The proportion should be some form of $\dfrac{4}{12} = \dfrac{4x + 5}{39}$; the solution is $x = 2$.

(d) The proportion should be some form of $\dfrac{i}{\frac{1}{4}i} = \dfrac{d^2}{3^2}$, which is equivalent to $\dfrac{1}{\frac{1}{4}} = \dfrac{d^2}{9}$. The solution is $d = 6$ feet. (*Note:* The variable i is an example of a *dummy variable;* it has no bearing on the solution.)

While you may not have had the identical proportions given in the answers above, you should have been able to change your proportions to the ones given by interchanging the means and/or extremes.

8. The idea of proportion is used to illustrate the relationships that can exist between two variables. Another method of showing these relationships is known as *variation.* While there is no essential difference between proportion and variation, the language and the equations involved are somewhat different, and some problems are easier to work by using variation.

The relationship between the variables (a, b), as described in the definition of *direct variation* given in review item 8, can be written symbolically as $\dfrac{a}{b} = k$, where k represents the constant of variation.

Here are some examples to show how this definition is used.

Example 1: Using symbols, write the statement *x varies directly as y, with a constant variation of* 26.

Solution: By direct application of the information in the definition, we can write $\frac{x}{y} = 26$. Naturally, by applying the axiom of multiplication, we could change this to $x = 26y$. Both forms are acceptable and each has its own advantages for certain situations.

Example 2: If *x* varies directly as y^2, and $x = 6$ when $y = 4$, find the constant of variation *k*.

Solution: From our definition of direct variation and the information in the problem, we can write $\frac{x}{y^2} = k$. Since we know that $x = 6$ when $y = 4$, we can write $\frac{6}{4^2} = k$. Thus, $k = \frac{6}{16}$ or $\frac{3}{8}$

Example 3: If *a* varies directly as *b*, and $a = 12$ when $b = 3$, find *a* when $b = 16$.

Solution: By using our definition, we can write $\frac{a}{b} = k$. And since we know that $a = 12$ when $b = 3$, we can write $\frac{12}{3} = k$, or $k = 4$. The next step is to return to the equation $\frac{a}{b} = k$ and write $\frac{a}{16} = 4$, from which $a = 64$ when $b = 16$.

All these examples may appear to make the use of variation more difficult than it really is. ("Explanations" often have this effect.) In actual practice, you will find that you will use proportion for some problems and variation for others. Solve the following problems by using direct variation.

(a) Find the constant of variation *k* if *a* varies directly as *b*, and $a = 6$ when $b = 2$.

(b) If $r = 56$ when $s = 4$, and if *r* varies directly as *s*, find *r* when $s = 3$.

(c) If *x* varies directly as y^3, and if $x = 64$ when $y = 2$, find *x* when $y = 1$.

------ ---------

(a) $k = 3$; (b) $r = 42$; (c) $x = 8$

9. As you might suspect, just as there are both direct and inverse proportion, there are also both direct and inverse variation. In symbols, we can define inverse variation as follows:

> If $a \cdot b = k$, then a varies inversely as b, and k is the constant of variation.

The following examples should help you understand the use of inverse variation in solving problems.

Example 1:   Find k if a varies inversely as b, and if $a = 5$ when $b = 8$.

Solution: Since a varies inversely as b, we can use our definition to write $a \cdot b = k$. Also, we know that $a = 5$ when $b = 8$; so we can write $5 \cdot 8 = k$, or $k = 40$.

Example 2:   If x varies inversely as y, and if $x = 6$ when $y = 2$, find the value of x when $y = 24$.

Solution: Since x varies inversely as y, we can use our definition to write $x \cdot y = k$. And since $x = 6$ when $y = 2$, we can write $6 \cdot 2 = k$, or $k = 12$. Using this value of k, we now can write $x \cdot y = 12$; hence, $x \cdot 23 = 12$ is solved to find that $x = \frac{1}{2}$ when $y = 24$.

Solve the following.

(a) x varies inversely as y. If $x = 9$ when $y = 2$, find x when $y = 3$.

(b) p varies inversely as v. If $p = 30$ when $v = 10$, find p when $v = 15$.

(c)   A wire 0.3 inch in diameter has 14 ohms resistance. How much resistance would there be in a wire of the same material and length, if the resistance varies inversely as the square of the diameter, and the new wire has a diameter of 0.25 inch?

(a) $x \cdot y = 9 \cdot 2 = 18 = k$; hence, $x \cdot 3 = 18$, or $x = 6$ when $y = 3$;

(b) $p \cdot v = 30 \cdot 10 = 300\ k$; hence, $p \cdot 15 = 300$, or $p = 20$ when $v = 15$;

(c) $rd^2 = k$; hence, $14(0.3)^2 = 1.26 = k$; therefore, $r = \dfrac{k}{d^2}$, or

$r = \dfrac{1.26}{.0625} = 20.16$ ohms

10.   The language of variation allows us to discuss quite easily the relations among more than two variables. Two kinds of variation involving two or more variables are *joint variation* and *combined variation*.

As shown in review item 10, an example of joint variation is the equation $\dfrac{x}{abc} = k$. We would read this as "x varies jointly as a, b, and c." Notice that in this equation, x varies *directly* as any variable (a, b, or c) when the other two variables are held constant.

Combined variation is illustrated by the equation $\frac{xy}{z} = k$, where k is, once again, the constant of variation. Notice in this case that x varies directly as z and inversely as y. Similarly, y varies directly as z and inversely as x. The wording of the problem will tell you when to use combined variation.

Each of the following questions involves one of the types of variation we have discussed. Supply the missing words in each case. (*Note:* x is always the variable related to the other variables, and k is always the constant of variation.)

(a) $\frac{x}{y} = k$, x varies _____ as y

(b) $xy = k$, x varies _____ as y

(c) $\frac{x}{ab} = k$, x varies _____ as the _____ of a and b

(d) $\frac{xy}{z} = k$, x varies directly as _____and inversely as _____; this is an example of _____ variation.

(e) $\frac{xy^2}{ab} = k$, x varies jointly as _____ and inversely as _____

Express the constant of variation as an integer or a fraction in simplest form in each of the following problems. (Do not perform any additional steps.)

(f) $x = 7$ and $y = 14$ when x varies directly as y. $k =$ _____

(g) $x = 7$ and $y = 14$ when x varies inversely as y. $k =$ _____

(h) $x = 3$, $y = 7$, and $z = 6$ when x varies jointly as y and z. $k =$ _____

(i) $x = 6$, $y = 7$, and $z = 3$ when x varies directly as y and inversely as the cube of z. $k =$ _____

(j) $x = 12$, $y = 3$, and $z = 2$ when x varies jointly as y and the square of z. $k =$ _____

- - - - - - - - - - - - - - -

(a) directly;   (b) inversely;   (c) jointly, product;   (d) z, y, combined;
(e) a and b, y^2: (f) $k = \frac{1}{2}$;   (g) $k = \frac{1}{14}$;   (i) $k = \frac{162}{7}$;   (j) $k = 1$

11. When solving a variation problem, after the constant of variation k is found, we are ready to express the equation in the most convenient form to complete the problem. For example, let us consider problem (f) from reference item 10 above. We want to write the equation of variation in a way convenient for finding x. And since we know that $k = \dfrac{1}{2}$, and that this is a problem in *direct* variation, we can write at once $\dfrac{x}{y} = \dfrac{1}{2}$, or $x = \dfrac{1}{2}y$. Substituting any given value for y will yield the value of x.

Write the equation of variation you would use to find x in problems (g) through (j) in reference item 10 above. Each equation should express x as a function of the other variables in the problem.

(g) $x =$ _____   (i) $x =$ _____

(h) $x =$ _____   (j) $x =$ _____

- - - - - - - - - - - - - -

(g) Since $xy = k$ (inverse variation) and $k = 98$, we can write $xy = 98$; $x = \dfrac{98}{y}$.

(h) Since $\dfrac{x}{yz} = k$ (joint variation) and $k = \dfrac{1}{14}$, we can write $x = \dfrac{yz}{14}$.

(i) Since $\dfrac{xz^3}{y} = k$ (combined variation) and $k = \dfrac{162}{7}$, we can write

$\dfrac{xz^3}{y} = \dfrac{162}{7}$; $x = \dfrac{162y}{7z^3}$.

(j) Since $\dfrac{x}{yz^2} = k$ (joint variation) and $k = 1$, we can write $\dfrac{x}{yz^2} = 1$; $x = yz^2$.

12.   The final step in solving problems by the use of variation is to find the value of the variable in question. For an example of a complete solution by the method of variation, we will use the facts given in problem (f) from reference item 10, but restate them in a slightly different way.

Example: If $x = 7$ when $y = 14$, and it is known that x varies directly as y, find x when $y = 6$.

Solution: Step 1: Set up the equation involving x, y, and k:

$$\frac{x}{y} = k$$

Step 2: Find k: $k = \dfrac{7}{14} = \dfrac{1}{2}$

Step 3: Write the equation of variation that will express x as a function of y: $\dfrac{x}{y} = \dfrac{1}{2}$; $x = \dfrac{1}{2}y$

Step 4: Compute x using the value of 6 for y:

$$x = \frac{1}{2}(6) = 3$$

To learn how to use the language of variation to establish an equation, solve the following problems by the use of variation.

(a) If x varies inversely as y and $x = 12$ when $y = 3$, find x when $y = 12$.

(b) The relation between x and y is shown in the table below. Find x when $y = 48$.

x	3	2	4	...
y	8	12	6	...

(c) If the temperature remains constant, the resistance of a wire varies directly as the length of the wire and inversely as the square of its diameter. If a wire whose diameter is 10 mils has a resistance of 0.3 ohms when its length is 400 feet, what is the resistance of a wire whose diameter is 15 mils and whose length is 2000 feet if the second wire is made from the same material as the first?

(d) If y varies directly as x and if the graph of $y = f(x)$ passes through the point (6, 18), find the abscissa if the ordinate is 24.

------ ----------

(a) Since $xy = k$ and we know that one pair of (x,y) values is (12, 3), we can find $k = 36$. Therefore, $x = 3$ when $y = 12$.

(b) You should observe that $x \cdot y$ is always 24; therefore we have an instance of inverse variation. If $xy = 24$, then $x = \dfrac{24}{48}$ or $x = \dfrac{1}{2}$ when $y = 48$.

(c) The equation would be $\dfrac{rd^2}{1} = k$ (where r = resistance, l = length, and d = diameter). We find $k = \dfrac{0.3}{4}$ or 0.075. Therefore, the resistance of the 2000-foot wire would be 0.666 ... ohms.

(d) We have $\dfrac{y}{x} = k$: so $k = 3$ and x (the abscissa) is 8 if y (the ordinate) is 24.

UNIT TWELVE

Solving Everyday Problems

Review Item	Ref Page	Example
1. *Algebraic representation* is choosing a letter of the alphabet to represent a number, usually the unknown quantity. *Translating* is turning words into mathematical symbols.	202	If the sum of two numbers is 7, and the letter n represents one of the numbers, the other number can be represented by $7 - n$. Five times a number equals 8 plus the number: $5n = 8 + n$
2. A *mathematical model* is an algebraic representation of a verbal statement. It is therefore, the result of translating.	203	In the example above, $5n = 8 + n$ is a mathematical model, the result of translating the word statement into symbols.
3. The four steps in solving word problems are: *Step 1: Representation* *Step 2: Translation* *Step 3: Solution* *Step 4: Check*	204	Twice a certain number increased by 12 equals 26. Let n = number Twice a number: $\qquad$ $2n$ increased by 12: $\qquad$ $+12$ equals 26: $\qquad\qquad$ $= 26$ or: $\qquad$ $2n + 12$ $\quad = 26$ $2n = 26 - 12 = 14; n \quad = 7$ Twice a number: $\quad 2 \cdot 7 = 14$ increased by 12: $\qquad\qquad +12$ equals 26: $\qquad\qquad\quad = 26$ or: $\qquad\qquad 14 + 12 = 26$? $\qquad\qquad\qquad\qquad 26 = 26$

Review Item	Ref Page	Example
4. Number problems containing two unknown quantities can be solved by either of the following methods:	205	One number is 3 more than twice the other. What are the numbers if their sum is 18?
a. Use some letter to represent one of the unknowns. Use one of the given relationships between them to express the other unknown in terms of the same letter. Use the second relationship to form an equation involving both unknowns. Then solve for both unknowns.		Let $s =$ smaller number; then $2s + 3 =$ larger number; hence, $s + (2s + 3) = 18$ and $s = 5$, $2s + 3 = 13$.
b. Represent each unknown by a different letter. Based on the two relationships described, write two separate equations. Solve them by the algebraic method for systems of equations which we reviewed in Unit 8.		Let $s =$ smaller number; let $l =$ larger number; then $2s + 3 = 1$. and $s + l = 18$, or $s + (2s + 3) = 18$ from which $s = 5$, and $l = 13$.
5. *Age problems:* two useful rules to remember: *Rule 1:* To find a person's future age, add the given number of years to his present age. *Rule 2:* To find a person's past age, subtact the given number of years from his present age.	206	In 3 years, a person x years old will be 21 years old: $x + 3 = 21$ Three years ago, a person x years old was 15 years old: $x - 3 = 15$
6. To represent a person's age on the basis of some past or future age, first arrive at an expression for his present age, and then add or subtact years as required.	207	To represent the age of a person 3 years hence if he was 16 years old 8 years ago: $(16 + 8) + 3 = 27$ years

Review Item	Ref Page	Example
7. *Percentage problems:* Percent means hundredths. A percent of a certain number is called a *percentage.* The number of which the percent is taken is the *base;* the percent itself is the *rate.*	209	25 is 10% $(\frac{10}{100})$ of 250. 250 is the base; 10% is the rate.
8. There are three kinds of percentage problems: a. Finding a percent of a number b. Finding what percent one number is of another c. Finding a number when a percent of it is known	209	What is 5% of 240? Let x = required number $= 0.05 (240)$ $= 12$ 7 is what percent of 34? Let x = required % $17 = \frac{x}{100} (34)$ $x = 50\%$ 86 is 40% of what number? Let x = required number $0.40x = 86$ $x = 215$
9. *Motion problems:* The three elements in a motion problem are: time T, speed or rate R, and distance D. They are related by the formulas $D = RT, R = \frac{D}{T}$ and $T = \frac{D}{R}.$ When working motion problems, keep in mind the following: Speed, rate, and velocity mean *uniform* speed or average speed. Units used to express time, speed, and distance must be consistent. The units usually used are hours, miles per hour, and miles. Divide minutes by 60 to convert minutes to decimal fractions of an hour.	211	A car traveling 60 mph will go 240 miles in 4 hours. $60 = \frac{240}{4}, 240 = 60 \cdot 4,$ $4 = \frac{240}{60}$

Review Item	Ref Page	Example
10. In a motion problem involving a *separation situation,* two travelers start from the same place at the same time and travel in opposite directions.	212	See reference item 10.
11. In motion problems involving a *closure situation,* two travelers start from separate points at the same time and travel toward each other until they meet.	213	See reference item 11.
12. In motion problems involving a *round-trip situation,* a traveler travels out and back to the starting place along the same road.	214	See reference item 12.
13. In motion problems involving the *gain* or *overtake situation,* after a traveler has begun his trip, a second traveler starts from the same place and, going in the same direction, overtakes the first.	215	See reference item 13.
14. Motion problems that require a solution for two unknowns and that contain enough information for two equations can be solved most easily by the method of simultaneous equations.	216	See reference item 14.

Review Item	Ref Page	Example
15. In working *mixture problems,* keep in mind that the total value of a number of units of the same kind is equal to the number of units times the value of one unit. To develop your equation for a mixture problem, find the value of each component of the mixture, and set the sum of these equal to the total value of the mixture.	218	See reference item 15.

UNIT TWELVE REFERENCES

1. The first difficulty—and sometimes the only difficulty—that many people have in working with word problems is that of turning words into mathematical symbols. Most of the words we use to express the relations between various quantities are associated with a mathematical operation. The most common of these have been assembled in the table below.

+	−	X	÷	=	*x* or another letter
sum and plus added to increased by more than	minus subtract less less than difference remainder decreased exceeds by	times product multiplied by	quotient divided by ratio	equals is is equal was equal is as much as	an unknown number

Refer to this chart as often as you need to in working with the concepts and problems throughout the rest of this unit.

To *translate* words into symbols accurately, you must be sure you know what the words mean. The way writers of textbooks or tests use words to express relationships between quantities may sometimes appear confusing. The best way to avoid confusion is to learn the meaning of the most commonly used expressions. Here are examples of three such expressions from the table above.

Increased by means *add*. Thus, *8 increased by 3 equals 11* may be written as $8 + 3 = 11$. Similarly, *decreased by* and *diminished by* both mean *subtract*. Thus, *9 decreased by 4 is 5* may be written as $9 - 4 = 5$.

Another way to indicate subtraction is by use of the word *exceeds*. Thus, *5 exceeds 3 by 2* means 5 is 3 more than 2, $5 - 3 = 2$, or $5 - 2 = 3$.

The following exercises include both addition and subtraction operations expressed in various ways. Read them carefully, and think about what they mean before you try to translate the word statements into algebraic equations. Where possible, solve the equations and check your solutions by substituting them back in the *original word statements*.

(a) Eight exceeds 1 by 7. _____

(b) *n* exceeds 6 by 7. _____ $n = $ _____

(c) Fifteen exceeds a certain number by 5. _____ $n = $ _____

(d) Six increased by 7 equals 17 decreased by 4. _____

(e) A number added to half itself equals 30. _____ $n = $ _____

(f) A number diminished by 4 equals the sum of 3 and 6. _____

  $n = $ _____

(g) The sum of 5 and 8 exceeds a certain number by 2. _____
  $n = $ _____

(h) *n* increased by 9 exceeds 7 by 10. _____ $n = $ _____

------ ---------
(a) $8 - 1 = 7$;   (b) $n - 6 = 7$, $n = 13$;   (c) $15 - n = 5$, $n = 10$;

(d) $6 + 7 = 17 - 4$;   (e) $n + \dfrac{n}{2} = 30$, $n = 20$;

(f) $n - 4 = 3 + 6$, $n = 13$;   (g) $(5 + 8) - n = 2$, $n = 11$;

(h) $(n + 9) - 7 = 10$, $n = 8$

2. We often refer to the algebraic representation of a verbal statement as a *mathematical model*. Below are some further examples to help familiarize you with the meaning of certain operational words and the way they are used. Note how the verbal statements have been translated into mathematical models.

Word Statement	Mathematical Model
Five *increased by* three means three *added to* five.	$5 + 3 = 8$
Half of ten *is* five means half of ten *equals* five.	$\dfrac{10}{2} = 5$
Twelve *decreased by* five means five *subtracted from* twelve.	$12 - 5 = 7$

Seven *exceeds* five by two.	$7 - 5 = 2$
Four *times* five is twenty.	$4 \cdot 5 = 20$
The positive *difference* of eight and three	$8 - 3 = 5$
if five.	

Here is some more practice. Translate each of the following statements into an equation. Then solve the equation. Use n for each unknown.

(a) If a number is decreased by 7, the result is 8.

(b) Four times a number, increased by 3, equals 19.

(c) Three-eighths of a number equals 12.

(d) A number added to one-third of itself equals 24.

(e) When four times a number is diminished by 10, the remainder is 26.

(f) Seven exceeds one-half a number by 4.

- - - - - - - - - - - - - - -

(a) $n - 7 = 8$, $n = 15$;   (b) $4n + 3 = 19$, $n = 4$;

(c) $\dfrac{3n}{8} = 12$, $n = 32$;   (d) $n + \dfrac{n}{3} = 24$, $n = 18$;   (e) $4n - 10 = 26$, $n = 9$;

(f) $7 - \dfrac{n}{2} = 4$, $n = 6$

3. Now let us look at the specific steps involved in solving word problems. The four essential steps of problem solving are as follows.

Step 1: *Representation* of the unknown(s). To represent the unknowns correctly, you must first read the problem carefully. Study it until the situation described is clear to you, and identify the quantities, both known and unknown, that are involved in the problem. Then select one of the unknowns and assign it a letter symbol, such as x. Express the other unknown(s) in terms of this letter, if you wish to use just one equation. *Remember:* if you decide to use more than one letter to represent the two unknowns, you will need to develop at least as many equations as there are unknowns.

Step 2: *Translation* of the relationships about the unknowns into an equation or system of equations. Search the problem for the information that tells you which quantities, or which combinations of quantities, are equal. When the desired combinations have been found, set them equal to each other, thus obtaining an equation.

Step 3: *Solution* of the equation or system of equations to find values of the unknowns.

Step 4: *Checking* of the values found to make sure they satisfy the original problem. Do *not* check your values in *your* equation, since the equation itself may be wrong! Always check the values you obtain in the original verbal statement.

Each of the following problems has one unknown. Solve.

(a) A certain number is equal to 35 decreased by 21. What is the number?

(b) 13 is equal to a number increased by 7. What is the number?

(c) The sum of a number and 11 is 23. What is the number?

(d) Seven added to three times a number is equal to 22. What is the number?

(e) Four times a certain number is equal to 35 decreased by the number. What is the number?

(f) If 25 is subtracted from a certain number, the difference is one-half the number. What is the number?

(g) Four times a certain number, decreased by 5, equals 25 diminished by 6 times the number. What is the number?

- - - - - - - - - - - - - - -

(a) $n = 35 - 21$, $n = 14$

check: $14 \overset{?}{=} 35 - 21$, $14 \overset{\checkmark}{=} 14$

(b) $13 = n + 7$, $n = 6$

(c) $n + 11 = 23$, $n = 12$

check: $12 + 11 \overset{?}{=} 23$, $23 \overset{\checkmark}{=} 23$

(d) $7 + 3n = 22$, $3n = 15$, $n = 5$

(e) $4n = 35 - n$, $5n = 35$, $n = 7$

check: $28 \overset{?}{=} 35 - 7$, $28 \overset{\checkmark}{=} 28$

(f) $n - 25 = \dfrac{n}{2}$, $n = 50$

(g) $4n - 5 = 25 - 6n$, $n = 3$

check: $12 - 5 \overset{?}{=} 25 - 18$, $7 \overset{\checkmark}{=} 7$

4. Review item 4 contains examples of the two methods for solving number problems in two unknowns. Below are some more representative problems of this kind. Solve them by either of the methods illustrated in the review item, following the four steps described in reference item 3.

(a) The sum of two numbers is 13 and their difference is 5. What are the two numbers?

(b) The larger of two numbers is equal to twice the smaller number increased by five. The smaller number equals the larger number decreased by 16. What are the numbers?

(c) Five times the smaller of two numbers decreased by seven is equal to twice the larger number increased by two. If the sum of the two numbers is 13, what are the numbers?

(d) The smaller of two numbers is equal to two less than half the larger number. If the larger number, increased by four, is equal to six times the smaller, what are the two numbers?

(e) The difference between two numbers is ten. Four times the larger decreased by 30 equals three times the smaller increased by twenty. What are the numbers?

(f) Three times the smaller of two numbers is equal to twice the larger. If the larger number plus two equals twice the smaller number less four, what are the numbers?

------ ---------

(a) $s = 4$, $l = 9$;   (b) $s = 11$, $l = 27$;   (c) $s = 5$, $l = 8$; (d) $s = 2$, $l = 8$;

(e) $s = 10$, $l = 20$;   (f) $s = 12$, $l = 18$

5. Follow the two rules given in review item 5 to represent the age of the person in each of the following problems.

Example:

(a) Ten years hence, if his present age is 25. *Answer:* $25 + 10$

(b) Ten years ago, if his present age is x years. _____

(c) Fifteen years hence, if his present age is y years. _____

(d) In x years, if his present age is 35 years. _____

(e) In y years, if his present age is x years. _____

(f) x years ago, if his present age is 50 years. _____

(g) y years ago, if his present age is k years. _____

------- ---------

(b) $(x - 10)$ years;   (c) $(y + 15)$ years;   (d) $(35 + x)$ years;   (e) $(x + y)$ years;
(f) $(50 - x)$ years;   (g) $(k - y)$ years

6.   In the problems for reference item 5, you were required to represent a person's age at another time on the basis of his present age. Conversely, age problems often require that you represent a person's present age on the basis of some past or future age.

For example, you might be asked to represent the age of a person 8 years hence, if he was 15 years old 5 years ago. Always start by arriving at an expression for his present age; then add or subtract years as described in the problem. In this case, you would represent his present age as $(15 + 5)$, to which you would then add 8 years to arrive at this age years from now. This gives us $(15 + 5) + 8 = 28$.

Now let us consider some typical age problems and how to go about solving them.

Example 1:   A man is 9 times as old as his son. In 3 years, the father will be only 5 times as old as his son. What is the present age of each?

Solution: Let x son's age now

$$9x = \text{Father's age now}$$
$$x + 3 = \text{son's age 3 years hence}$$
$$9x + 3 = \text{father's age 3 years hence}$$

From the statement of the problem:

$$9x + 3 = 5(x + 3)$$
$$9x + 3 = 5x + 15$$
$$9x - 5x = 15 - 3$$
$$4x = 12$$
$$x = 3 \text{ (son's age now)}$$
$$9x = 27 \text{ (father's age now)}$$

Check: $9(3) + 3 \overset{?}{=} 5(3 + 3)$; $30 \overset{\checkmark}{=} 30$

Example 2:   Bill is 10 years older than Ruth. In 8 years, twice Bill's age will equal 3 times Ruth's age. What are their present ages?

Solution: Let $x = $ Ruth's age now

$$x + 10 = \text{Bill's age now}$$
$$x + 8 = \text{Ruth's age 8 years hence}$$
$$x + 18 = \text{Bill's age 8 years hence}$$
$$\text{then } 2(x + 18) = 3(x + 8)$$
$$2x + 36 = 3x + 24$$
$$x = 12 \text{ (Ruth's age now)}$$
$$x + 10 = 22 \text{ (Bill's age now)}$$

In the examples above, we made use of the fact that the difference between two people's ages remains constant during their lifetimes.

Use these examples to guide you in solving the following problems.

(a) Walter is 8 years older than Ray. In 6 years, 5 times Walter's age will equal 9 times Ray's age. How old is each at present?

(b) The sum of the ages of Mary and her mother is 60 years. In 20 years, twice Mary's age increased by her mother's age (then) will equal 138 years. How old is each now?

(c) Robert is 14 years old and his father is 38 years old. How many years ago was the father exactly seven times as old as his son? (*Hint:* Let x = the number of years ago that the father's age was 7 times his son's age. Start by representing both son's and father's previous ages in terms of their present ages minus x.)

(d) Two years ago, a woman was four times as old as her son. Three years from now the mother will be only 3 times as old as the son. How is each at present? (*Hint:* Let x and $4x$ represent their ages 2 years ago.)

(e) A man was 30 years of age when his daughter was born. The father's age now exceeds 3 times the daughter's age by 6 years. How old is each at present?

(f) Fred is four times as old as Cliff. In 10 years, he will be only twice as old as Cliff is then. How old is each?

------- ----------

(a)   Let x = Ray's age now
        $x + 8$ = Walter's age now
        $x + 6$ = Ray's age 6 years hence
        $x + 14$ = Walter's age 6 years hence
    Then $5(x + 14) = 9(x + 6)$
        $5x + 70 = 9x + 54$
            $4x = 16$
            $x = 4$ (Ray's age now)
        $x + 8 = 12$ (Walter's age now)

(b) x and $60 - x$ represent their ages now. $x + 20$ and $80 - x$ represent their ages 20 years hence. Equation is $2(x + 20) + (80 - x) = 138$ from which $x = 18$ and $60-x = 42$, their present ages.

(c) Let x = number of years ago father's age was 7 times the son's age
            $14 - x$ =   son's age then
            $38 - x$ =   father's age then
        Equation: $(38 - x) = 7(14 - x)$
                $x = 10$ years

(d)   Equation: $4x + 5 = 3(x + 5)$
$$x = \quad 10 \text{ (son's age two years ago)}$$
$$x + 2 = \quad 12 \text{ (son's age now)}$$
$$4x + 2 = \quad 42 \text{ (mother's age now)}$$

(e)   Let $x =$ daughter's age now
$30 + x =$ father's age now
Equation: $(30 + x) - 6 = 3x$
$$x = \quad 12$$
$$30 + x = \quad 42$$

(f)   Let $x =$ Cliff's age now
$$4x = \quad \text{Fred's age now}$$
$$x + 10 = \quad \text{Cliff's age 10 years hence}$$
$$4x + 10 = \quad \text{Fred's age 10 years hence}$$
Equation: $4x + 10 = 2(x + 10)$
$$x = \quad 5 \text{ (Cliff's age now)}$$
$$4x = \quad 20 \text{ (Fred's age now)}$$

7. As explained in review item 7, a *percent* of a certain number is called a *percentage P*. The number of which the percent is taken is called the *base B*. The percent itself is called the *rate R*. Percent means hundredths; it is abbreviated %. When solving percentage problems, always keep in mind that percentage is simply a common way of expressing a fraction whose denominator is 100. The familiar relationship among these three quantities is $P = RB$.

People who try to work percentage problems arithmetically often become confused, because they are not sure what operations are to be performed with the numbers. Algebra simplifies the solution of such problems.

For a review of the procedures involving percentages that you learned in arithmetic, solve the following problems.

(a) What is 5% of 240?_____

(b) 34 is what percent of 85?_____

(c) 117 is 65% of what number?_____

------ ---------

(a) 12;   (b) 40%;   (c) 180

8. If you got the correct answers to the problems in reference item 7, you remember the proper procedures involving percentages from arithmetic. This will help you in comparing those procedures with the ones used in algebra.

The algebraic method of solving percentage problems, characterized by the familiar starting point, "Let $x = \cdots$" is illustrated in the three examples shown in review item 8. Each shows the proper procedure to follow for one of the three basic types of percentage problems.

Below you will find an assortment of percentage problems. It is up to you to select and apply the correct procedures for solving them.

(a) 3% of $500

(b) 35% of $200

(c) 7% of $800

(d) 7 is what percent of 14?

(e) 15 is what percent of 25?

(f) 60 is what percent of 40?

(g) 20 is 40% of what number?

(h) 12 is what percent of 16?

(i) 12 is 12% of what number?

(j) 68 is 200% of what number?

(k) What is $12\frac{1}{2}$% of 96?

(l) A baseball team won 9 games and lost 3. What percent of its games did it win?

(m) A man received $24 on an investment of $400. What rate did he receive?

(n) A merchant bought a chair for $100 and marked the price up 40% for resale. What price did he put on it?

(o) What is 0.1% of $5000?

------ ---------
(a) $15; (b) $70; (c) $56; (d) 50%; (e) 60%; (f) 150%; (g) 50; (h) 75%; (i) 100; (j) 34; (k) 12; (l)75%; (m) 6%; (n) $140; (o) $5

A slightly different and more complex variety of percentage problem appears below.

Example: 10% of a number increased by 28% of the same number equals 57. What is the number?

Solution: Let $x =$ the number

$$0.10x + 0.28x = 57$$
$$0.38x = 57$$
$$38x = 5700$$
$$x = 150$$

Check: $0.10(150) + 0.28(150) \overset{?}{=} 57; 57 \overset{\checkmark}{=} 57$

Use this approach to solve the following problems.

(a) 8% of a number plus 12% of the number is 62. What is the number?

(b) A number decreased by 10% of itself equals 405. What is the number?

(c) 10% of a number plus 8% of the number, decreased by 6% of the number, equals 42. What is the number?

(d) A radio was sold for $68.00 after discounts of 10% and 5% off the list price were allowed. What was the list price of the radio?

(e) A baseball team with a winning percentage of 0.839 has won 94 games. How many games has it lost? (*Hint:* Let $x =$ total number of games played.)

(f) How cheaply can a grocer afford to sell berries that cost 12¢ a quart if he must make a profit of 20% based on the selling price? (*Hint:* Let $x =$ selling price in cents.)

- - - - - - - - - - - - - - -

(a) $0.08x + 0.12x = 62$; $x = 310$, the number;
(b) $x - 0.10x = 405$; $0.9x = 405$; $x = 450$, the number;
(c) $0.10x + 0.08x - 0.06x = 42$; $0.12x = 42$; $x = 350$;
(d) $x - 0.10x - 0.05x = \$68$; $0.85x = \$68$; $x = \$80$, the list price of the radio;
(e) $0.839x = 94$; $x = 112$, total games played; $112 - 94 = 18$, number of games lost;
(f) $x - 0.20x = 12¢$; $0.8x = 12¢$; $x = 15¢$, selling price

9. Now it is time to talk about *motion problems,* of which there are several types. As stated in review item 9, the three elements common to motion problems are time T, rate R, and distance D. You will recall that we worked with these quantities in Unit 7 and found that they are related as follows:

$$D = RT \qquad R = \frac{D}{T} \qquad T = \frac{D}{R}$$

For practice using these formulas, solve the following problems.

(a)   A car travels at a speed of 60 mph for 3 hours. How far does it go?

(b)   An airplane flies 900 miles in 2 hours. What is its average rate of speed?

(c)   A train travels 250 miles at an average speed of 50 mph. How long does it take to make the trip?

- - - - - - - - - - - - - -

(a) (60 mph) (3 hours) = 180 miles;   (b) $\dfrac{900 \text{ miles}}{2 \text{ hours}} = 450 \text{ mph}$;

(c) $\dfrac{250 \text{ miles}}{50 \text{ mph}} = 5 \text{ hours}$

10.   Let us try solving some separation problems that require the application of the time-speed-distance formula.

Example 1:   Suppose you get in your car and drive east at 40 mph. At the same time you leave, a friend gets in his car and drives west, leaving from the same point, at 50 mph. How far apart will the two of you be at the end of 4 hours?

Solution: Since motion problems are easier to visualize—and therefore to solve—with the aid of a diagram, we have illustrated the situation below.

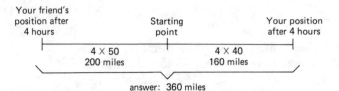

answer: 360 miles

Example 2:   Two planes leave the same airport at the same time and fly in opposite directions. The speed of the faster plane is 100 mph faster than that of the slower plane. At the end of 5 hours, they are 2000 miles apart. Find the rate of each plane.

Solution: Let R =   rate of slower plane
$R + 100$ =   rate of faster plane
$5R + 5(R + 100)$ =   2000 (since the sum of the distances each flew in 5 hours equals 2000 miles)
R =   150 mph (rate of slower plane)
$R + 100$ =   250 mph (rate of faster plane)

The kind of problem illustrated in these two examples is known as a *separation situation* or *separation problem*. Below are three more problems of this type. In each of these problems, two people start from the same place at the same time and travel in opposite directions. Draw a simple diagram of the situation in each case to assist you in visualizing it. Use R for rate, T for time, and D for distance. Then develop a mathematical model for each problem.

(a) After 3 hours two drivers are 330 miles apart; one is traveling 10 mph faster than the other. Find the rate of speed of the slower driver.

(b) At speeds of 40 mph and 20 mph two cars travel the same amount of time until they are 420 miles apart. How many hours do they travel?

(c) At speeds in the ratio of 7:3, two drivers are 360 miles apart at the end of 3 hours. Find the rate of each. (*Hint:* Use $7R$ and $3R$ to represent their respective rates.)

- - - - - - - - - - - - - -

(a) $3R + 3(R + 10) = 330$; $R = 50$ mph;

(b) $40T + 20T = 420$; $T = 7$ hours;

(c) $3(7R) + 3(3R) = 360$; $R = 12$ mph (faster rate is 84 mph; slower rate is 36 mph)

11. In the *closure situation,* the travelers are moving toward each other, not away from each other as in the separation situation.

Example: Jean and Sharon, who are 568 miles apart, start driving toward each other in their cars. Jean drives 40 mph, and Sharon drives 36 mph. If Jean has an accident and is delayed 1 hour before continuing her trip, how soon will they meet?

Solution: What we are seeking here is time T, the interval between the instant at which they start and the moment they finally meet. What effect does Jean's delay have on this? It simply requires us to represent her time by $(T - 1)$. Therefore, we can state the following:

$$\text{Distance Sharon travels} = 36T$$
$$\text{Distance Jean travels} = 40(T - 1)$$
$$36t + 40(T - 1) = 568$$
$$76T - 40 \quad = 568$$
$$T \quad = 8 \text{ hours}$$

Here are some similar problems for you to solve.

(a) Two trains start at the same time for towns 385 miles apart, and meet in 5 hours. If the rate of one train is 7 mph less than the rate of the other train, what is the rate of each?

(b) Two cars start at the same time from towns 448 miles apart, and meet in 4 hours. If the rate of one is 8 mph more than the other, what is the rate of each?

(c) Starting 297 miles apart, two drivers travel toward each other at rates of 38 mph and 28 mph until they meet. What is the travel time of each?

- - - - - - - - - - - - - - -

(a) Let x = rate of faster train

$$x - 7 \ = \text{rate of slower train}$$
$$5x + 5(x - 7) = 385$$
$$x \ = 42 \text{ mph (faster train)}$$
$$x - 7 \ = 35 \text{ mph (slower train)}$$

(b) 52 and 60 mph; (c) $4\frac{1}{2}$ hours

12. Below is a typical *round-trip situation*.

Example: Paul drove from his home to Boston and back again along the same road in 10 hours. His average speed going was 20 mph, and his average speed returning was 30 mph (obviously, he was not on an expressway!). How long did Paul take in each direction, and what distance did he cover each way?

Solution: Let T = time going

$$10 - T \ = \text{ time returning}$$
$$20T \ = \text{ distance going}$$
$$30(10 - T) \ = \text{ distance returning}$$
$$20T \ = 30(10 - T)$$
$$T \ = 6 \text{ hours (time going)}$$
$$10 - T \ = 4 \text{ hours (time returning)}$$

Both $20T$ and $30(10 - T) = 120$ miles (distance each way)

Using this same approach, solve the following problems.

(a) A traveler took 2 hours more returning from a trip than going. He averaged 50 mph out and 45 mph back. Find the time going.

(b) Mrs. Goldsby drove her car from her home to Ventura at a rate of 35 mph, and returned at a rate of 40 mph. Find her time going and returning, if the time returning was 1 hour less than the time going.

(c) A traveler takes 3 hours less to travel back to his starting point than he did to go out. If he averaged 45 mph out and 54 mph back, find his time to return.

(d) After taking 3 hours to go out from his starting place, a traveler returns in 5 hours, averaging 28 mph slower on the way back. What was his average rate going?

------ ---------

(a) $50t = 45(T + 2)$, $T = 18$ hours (time going);

(b) $T =$ time going; $35T = 40(T - 1)$; $T = 8$ hours (time going); $T - 1 = 7$ hours (time returning);

(c) 15 hours; (d) 70 mph

13. Below is an example of the *overtake situation*, a type of motion problem frequently encountered.

Example: An hour after Mike left on a week-end bicycle trip, his family noticed that he had forgotten to take his sleeping bag. His brother Ed jumped into his car and started after Mike. If Mike was traveling at the rate of 8 mph and Ed drove at the rate of 40 mph, how long did it take Ed to catch up with Mike?

Solution: As in the case of the round-trip situation, the key to this kind of problem is that the distance traveled by each individual is the same. Therefore, we start with the idea of equating the distances traveled by the two brothers. The *times* traveled will, of course, be different, but the distances will not.

$$\text{Let } x = \text{Ed's time}$$
$$x + 1 = \text{Mike's time (since he started an hour before his brother)}$$
$$40x = 8(x + 1) \text{ (since they traveled equal distances)}$$
$$x = \frac{1}{4} \text{ hour (the time it took Ed to catch up with Mike)}$$

Use this approach to solve the following overtake problems.

(a) A traveler begins a trip traveling at a rate of 20 mph. Three hours later a second traveler, proceeding at 40 mph, sets out to overtake him. How long will it take him to do so?

(b) Two and one-half hours after a traveler has begun his trip, a second traveler starts from the same place and overtakes him after traveling 6 hours at a rate of 34 mph. Find the rate of the first traveler.

(c) A traveler moves 18 mph slower than twice the speed of a second traveler. If the second traveler starts out 2 hours later than the first traveler and overtakes him in 5 hours, find the rate of the second traveler.

------ ---------

(a) Let T = time of second traveler
    $T + 3$ = time of first traveler
    $40T$ = $20(T + 3)$
       T = 3 hours
(b) 24 mph: (c) 14 mph

14. Learn to recognize motion problems that require a solution for two unknowns and that include enough information for two equations. In such cases, solution by the method of simultaneous equations is usually easiest. Here is one such problem.

Example: Dr. Conover covered 310 miles by traveling for 4 hours at one speed and then for 5 hours at another speed. Had she gone 5 hours at the first speed and 4 hours at the second speed, she would have covered 320 miles. Find the two speeds.

Solution: Obviously, the two speeds are our unknowns. Therefore, if we let x = the first speed and y = the second speed, we get:

$$(1) \quad 4x + 5y = 310$$
$$(2) \quad 5x + 4y = 320$$

Multiplying (1) by 5: $\quad 20x + 25y = 1550$
Multiplying (2) by 4: $\quad 20x + 16y = 1280$

Subtracting (2) from (1): $\quad 9y = 270$
$$y = 30$$
Substituting $y = 30$ in (1): $\quad 4x + 5 \cdot 30 = 310$
$$x = 40 \text{ mph}$$

Solve the following using the above procedure.

(a) By going 20 mph for one period of time and 30 mph for another, Mr. Smith traveled 280 miles. Had he gone 10 mph faster in each case, he would have covered 390 miles. How long did he travel at each speed?

(b) By traveling for 2 hours at one speed and then 7 hours at another, Dianne completed a trip of 258 miles. Had her first rate been 10 mph faster and her second rate twice as fast, she would have gone 488 miles. What were her two rates of speed?

------ ---------

(a) This problem differs from the example only in that the two unknown values are time (hours) rather than speed.

$$\text{Let } x = \text{first time (at 20 mph)}$$
$$y = \text{second time (at 30 mph)}$$
$$(1)\ 20x + 30y = 280$$
$$(2)\ 30x + 40y = 390$$
$$(10 \text{ mph faster in each case})$$

Multiplying (1) by 3: $60x + 90y = 840$
Multiplying (2) by 2: $60x + 80y = 780$

Subtracting (2) from (1): $10y = 60$
$$y = 6 \text{ hours}$$
$$x = 5 \text{ hours}$$

(b) 24 mph for 2 hours and 30 mph for 7 hours

Below are examples of each of the five kinds of motion problems we have reviewed. Solve each by referring to the appropriate review or reference item for help if you need it.

(a) An airplane traveled from its base to a distant point and back again along the same route in a total of 8 hours. Its average rate going was 180 mph, and its average rate returning was 300 mph. How long did it take in each direction and what was the distance covered each way?

(b) Two trains leave the same terminal at the same time and travel in opposite directions. After 8 hours they are 360 miles apart. The speed of the faster train is 3 mph less than twice that of the lower train. Find the rate of speed of each train.

(c) A boat travels at 24 mph. A patrol boat starts 3 hours later from the same place and travels at 32 mph in the same direction. How long will it take to overtake the first boat?

(d) Two planes leave from points 1925 miles apart at the same time and fly toward each other. Their average speeds are 225 mph and 325 mph, respectively. How soon will the planes meet?

(e) By traveling for 5 hours at one speed and then for 3 hours at another, Mr. Stewart covered 250 miles. Had he traveled for 2 hours longer at each speed, he would have covered 370 miles. Find the two rates at which he traveled.

------ ----------

(a) Time going was 5 hours; time returning was 3 hours; distance covered each way was 900 miles.

(b) Rates were 16 mph and 29 mph.

(c) Time needed to overtake is 9 hours.

(d) Planes will meet in $3\frac{1}{2}$ hours.

(e) Rates were 35 mph and 25 mph.

15.   Mixture problems can sometimes appear confusing and difficult to handle unless you have your approach pretty well in mind. To learn how to approach such problems, study the following example and then work a few problems yourself.

Example: A seedsman has clover seed worth 30¢ a pound and timothy seed worth 12¢ a pound. How many pounds of each should he use to make a mixture of 300 pounds worth 18¢ a pound?

Solution: The important point to bear in mind when working mixture problems is this: The total value of a number of units of the same kind equals the number of units times the value of one unit.

We know the value of the clover (30¢ per pound) and of the timothy seed (12¢ per pound). But since we do not know the weight of either, we will have to let x equal the weight of the clover seed and $(300 - x)$ equal the weight of the timothy seed, since together they form a mixture that will weigh 300 pounds.

$$
\begin{aligned}
30x &= \text{value of clover in cents}\\
12(300 - x) &= \text{value of timothy in cents}\\
18(300) &= \text{total value of mixture}\\
\text{Then } 30x + 12(300 - x) &= 18(300)\\
18x &= 1800\\
x &= 100 \text{ (pounds of clover)}\\
300 - x &= 200 \text{ (pounds of timothy)}
\end{aligned}
$$

Notice that we could have solved this problem using two unknowns. If we let y represent the weight of the timothy seed, we have as a second equation $x + y = 300$. This would allow a solution by simultaneous equations. The result, of course, would have been the same. Try it for yourself and see.

   Following are a few mixture problems for you to practice on. Some are quite similar to the example above; others are slightly different. They all can be solved by following the basic principles and procedures used in the example. Do not be alarmed or discouraged if each new problem you read sounds different from the preceding one. Remember that you are learning, or relearning, to apply basic principles and procedures.

(a) A dealer in tea has one brand worth 50¢ a pound and another worth 80¢ a pound. How many pounds of each kind must he use to make a 60 pound mixture worth 72¢ per pound?

(b) A coffee merchant blended coffee worth 93¢ a pound with coffee worth $1.20 a pound. The mixture of 30 pounds was valued by him at $1.02 a pound. How many pounds of each grade did he use?

(c) Maple syrup worth $6.00 a gallon and corn syrup worth 80¢ a gallon are used to make a mixture worth $2.36 a gallon. How many gallons of each kind of syrup are needed to make 50 gallons of the mixture?

(d) An order of candy cost $14. It contained one kind of candy worth 60¢ a pound and another worth 50¢ a pound. If there were 5 more pounds of the 60¢ candy, how many pounds of each kind were in the order? (*Hint:* Let x = pounds of kind worth 50¢, $x + 5$ = pounds of kind worth 60¢.)

------ ---------

(a) x = pounds of 50¢ tea, $60 - x$ = pounds of 80¢ tea; then $50x + 80$
    $(60 - x = 72(60)$; $x = 16$ pounds, $60 - x = 44$ pounds

(b) x = pounds of cheaper coffee, $30 - x$ = pounds of better coffee; then $93x + 120(30 - x) = 102(30)$; $x = 20$ pounds, $30 - x = 10$ pounds

(c) x = gallons of maple syrup, $50 - x$ = gallons of corn syrup; then $6x + .80(50 - x) = 2.36(50)$; $x = 15$ gallons, $50 - x = 35$ gallons

(d) x = pounds of 50¢ candy, $x + 5$ = pounds of 60¢ candy; then $.50x + 60(x + 5) = 14$; $x = 10$ pounds, $x + 5 = 15$ pounds

Appendix

GLOSSARY OF TERMS

Absolute value (of a signed number): The numerical value of a number without regard to its sign. Thus $|x| = x$, $x \geqslant 0$ and $|x| = -x$, $x < 0$.

Algebraic representation: Use of mathematical symbols to represent word statements.

Axiom: A statement that is accepted without proof.

Binomial: An algebraic expression of two terms.

Checking (an equation): Substituting solution values for the unknown back in the original equation (or statement of the problem) to verify that it reduces it to an identity.

Clearing (an expression of) fractions: Multiplying all terms in an expression by the LCD in order to establish a common denominator.

Coefficient: The name given to a factor (or group of factors) of a product to describe its relation to the remaining factors.

Common monomial factor: The combined literal and numerical factors common to all terms of a polynomial.

Cross product: The sum of the products of the inner and outer terms of a binomial product.

Denominator: The numbers (or letters) below the fraction bar.

Digit: Arabic symbol for a number.

Dividend: The number being divided.

Divisor: The number by which the dividend is being divided.

Equation: A statement of equality between two algebraic expressions.

Evaluate: Substituting numbers for letters to find the value of an algebraic expression.

Exponent: The small number written to the right of and slightly above another number to indicate how many times the latter is to be used as a factor.

Expression (algebraic): A collection of terms combined by symbols of operation and grouping.

Factor: Any one of two or more numbers or letters which when multiplied together form a product.

Factoring: Expressing a number (or algebraic expression) as a product of prime factors.

Formula: Use of symbols to express a rule in brief form.

Fraction: The quotient of two numbers or expressions; part of any object, quantity, or digit; usually indicated by a numerator and denominator, separated by a fraction bar.

Identity: An unconditional equality.

Integer: A whole number.

Inverse: Having the opposite effect. An inverse operation has the effect of undoing another operation. (For example, subtraction is the inverse of addition, and division is the inverse of multiplication.)

Inversely: Oppositely.

Inversely proportional: Increasing or decreasing oppositely.

Literal factor: A letter used as a factor.

Literal number: A letter used to represent a number.

Lowest common denominator (LCD): The smallest number to which all others of a group may be changed.

Minuend: A number from which another number is being subtracted.

Monomial: An algebraic expression of one term.

Negative number: A number whose value is less than zero.

Number: The concept of quantity, or count.

Number scale (or line): A scale along which equally spaced positive and negative numbers represent distances (in opposite directions) from zero.

Numeral: A symbol for a number; a digit from 0 to 9, or a combination of digits.

Numerator: The numerals or letters lying above a fraction bar.

Numerical factor: A numeral used as a factor.

Polynomial: An algebraic expression that has only positive whole numbers for the exponents of the variables.

Positive number: A number whose value is greater than zero.

Power (of a number): The product of a number used two or more times as a factor.

Prime number: A whole number greater than 1, that has no whole number factors except 1 and itself.

Product: The result of multiplication.

Proportional(to): Changing correspondingly with some other quantity or term.

Quotient: The result of division.

Reciprocal: The reciprocal of a number is 1 divided by that number.

Root (of an equation): A number which, when substituted for the unknown letter, makes the equation a true statement (that is, both sides of the equation equal).

Satisfy (an equation): Finding the numerical values for the literal terms of an equation that will reduce it to an identity or a true statement.

Signed number: A positive or negative number, preceded by a plus sign (or no sign) if it is positive, and by a minus sign if it is negative.

Signs of operation: Plus or minus signs used to indicate addition or subtraction.

Signs of quality: Plus or minus signs used to indicate positive or negative numbers.

Simplify: Combine like terms of an algebraic expression in order to condense it as much as possible.

Solving (an equation): Finding values of the unknown which when substituted in the equation will make it a true statement. (See *Root.*)

Square root: One of the two equal factors of a number.

Square root, principal: The positive square root of a number or algebraic expression.

Subtrahend: The number being subtracted.

Symbol (numerical): A figure that stands for a number. (See *Numeral.*)

Term: A number, letter, or the product of numbers and letters.

Terms, like: Terms that differ only in their numerical coefficients (that is, that have the same literal part).

Terms, unlike: Terms having different literal coefficients or whose literal factors have different exponents.

Trinomial: An algebraic expression of three terms.

Whole number: The set of counting numbers and zero.

SYMBOLS USED IN THIS BOOK

$\lvert a \rvert$	absolute value
$=$	equals
$>$	greater than
$<$	less than
$\geqslant$	greater than or equal to
$\leqslant$	less than or equal to
$\ldots$	and so on
$\sqrt{}$	square root
$+$	addition or positive
$-$	subtraction or negative
$\cdot, \times$	multiplication
$\div$	division
$\neq$	not equal to

$\ngtr$	not greater than
$\nless$	not less than
(x, y)	coordinates of a point
(a, b)	ordered pair
$:$	ratio

TABLE OF POWERS AND ROOTS

NO.	SQ.	SQ. ROOT	CUBE	CUBE ROOT	NO.	SQ.	SQ. ROOT	CUBE	CUBE ROOT
1	1	1.000	1	1.000	51	2,601	7.141	132,651	3.709
2	4	1.414	8	1.260	52	2,704	7.211	140,608	3.733
3	9	1.732	27	1.442	53	2,809	7.280	148,877	3.756
4	16	2.000	64	1.587	54	2,916	7.348	157,464	3.780
5	25	2.236	125	1.710	55	3,025	7.416	166,375	3.803
6	36	2.449	216	1.817	56	3,136	7.483	175,616	3.826
7	49	2.646	343	1.913	57	3,249	7.550	185,193	3.848
8	64	2.828	512	2.000	58	3,364	7.616	195,112	3.871
9	81	3.000	729	2.080	59	3,481	7.681	205,379	3.893
10	100	3.162	1,000	2.154	60	3,600	7.746	216,000	3.915
11	121	3.317	1,331	2.224	61	3,721	7.810	226,981	3.936
12	144	3.464	1,728	2.289	62	3,844	7.874	238,328	3.958
13	169	3.606	2,197	2.351	63	3,969	7.937	250,047	3.979
14	196	3.742	2,744	2.410	64	4,096	8.000	262,144	4.000
15	225	3.873	3,375	2.466	65	4,225	8.062	274,625	4.021
16	256	4.000	4,096	2.520	66	4,356	8.124	287,496	4.041
17	289	4.123	4,913	2.571	67	4,489	8.185	300,763	4.062
18	324	4.243	5,832	2.621	68	4,624	8.246	314,432	4.082
19	361	4.359	6,859	2.668	69	4,761	8.307	328,509	4.102
20	400	4.472	8,000	2.714	70	4,900	8.367	343,000	4.121
21	441	4.583	9,261	2.759	71	5,041	8.426	357,911	4.141
22	484	4.690	10,648	2.802	72	5,184	8.485	373,248	4.160
23	529	4.796	12,167	2.844	73	5,329	8.544	389,017	4.179
24	576	4.899	13,824	2.884	74	5,476	8.602	405,224	4.198
25	625	5.000	15,625	2.924	75	5,625	8.660	421,875	4.217
26	676	5.099	17,576	2.962	76	5,776	8.718	438,976	4.236
27	729	5.196	19,683	3.000	77	5,929	8.775	456,533	4.254
28	784	5.292	21,952	3.037	78	6,084	8.832	474,552	4.273
29	841	5.385	24,389	3.072	79	6,241	8.888	493,039	4.291
30	900	5.477	27,000	3.107	80	6,400	8.944	512,000	4.309
31	961	5.568	29,791	3.141	81	6,561	9.000	531,441	4.327
32	1,024	5.657	32,768	3.175	82	6,724	9.055	551,368	4.344
33	1,089	5.745	35,937	3.208	83	6,889	9.110	571,787	4.362
34	1,156	5.831	39,304	3.240	84	7,056	9.165	592,704	4.380
35	1,225	5.916	42,875	3.271	85	7,225	9.220	614,125	4.397
36	1,296	6.000	46,656	3.302	86	7,396	9.274	636,056	4.414
37	1,369	6.083	50,653	3.332	87	7,569	9.327	658,503	4.431
38	1,444	6.164	54,872	3.362	88	7,744	9.381	681,472	4.448
39	1,521	6.245	59,319	3.391	89	7,921	9.434	707,969	4.465
40	1,600	6.325	64,000	3.420	90	8,100	9.487	729,000	4.481

TABLE OF POWERS AND ROOTS

NO.	SQ.	SQ. ROOT	CUBE	CUBE ROOT	NO.	SQ.	SQ. ROOT	CUBE	CUBE ROOT
41	1,681	6.403	68,921	3.448	91	8,281	9.539	753,571	4.498
42	1,764	6.481	74,088	3.476	92	8,464	9.592	778,688	4.514
43	1,849	6.557	79,507	3.503	93	8,649	9.644	804,357	4.531
44	1,936	6.633	85,184	3.530	94	8,836	9.695	830,584	4,547
45	2,025	6.708	91,125	3.557	95	9,025	9.747	857,375	4.563
46	2,116	6.782	97,336	3.583	96	9,216	9.798	884,736	4.579
47	2,209	6.856	103,823	3.609	97	9,409	9.849	912,673	4.595
48	2,304	6.928	110,592	3.634	98	9,604	9.899	941,192	4.610
49	2,401	7.000	117,649	3.659	99	9,801	9.950	970,299	4.626
50	2,500	7.071	125,000	3.684	100	10,000	10.000	1,000,000	4.462

Index